AT EVERY DEPTH

AT EVERY DEPTH

Our Growing Knowledge of the Changing Oceans

TESSA HILL AND
ERIC SIMONS

Columbia University Press
New York

Columbia University Press
Publishers Since 1893
New York   Chichester, West Sussex
cup.columbia.edu

Copyright © 2024 Tessa M. Hill and Eric Simons
All rights reserved

Library of Congress Cataloging-in-Publication Data
Names: Hill, Tessa, author. | Simons, Eric, author.
Title: At every depth : our growing knowledge of the changing oceans / Tessa Hill and Eric Simons.
Description: New York : Columbia University Press, 2024. | Includes bibliographical references and index.
Identifiers: LCCN 2023031782 | ISBN 9780231199704 (hardback) | ISBN 9780231553254 (ebook)
Subjects: LCSH: Oceanography. | Marine ecology. | Human ecology.
Classification: LCC GC28 .H55 2024 | DDC 551.46—dc23/eng/20231012
LC record available at https://lccn.loc.gov/2023031782

Printed in the United States of America

Cover design: Noah Arlow and Julia Kushnirsky
Cover image: Shutterstock

We dedicate this to the observers, the recorders, the messengers, and those carrying transect tapes into the cold and dark to illuminate our world. May your observations provide clarity for our future.

For Lawson, Cormac, and Brian
All of my songs and stories are for you
—T. M. H.

For Hari, Eleanor, and Margaret
My rocks in the changing sea
—E. S.

CONTENTS

Prologue   ix

1 The Tide Pool   1
2 The Reef   23
3 The Forest   53
4 The Gardens   75
5 The Abundant Ocean   97
6 The Open Ocean   131
7 The Polar Worlds   149
8 The Deep   173

Epilogue   209

Acknowledgments   215

Notes   217

Index   257

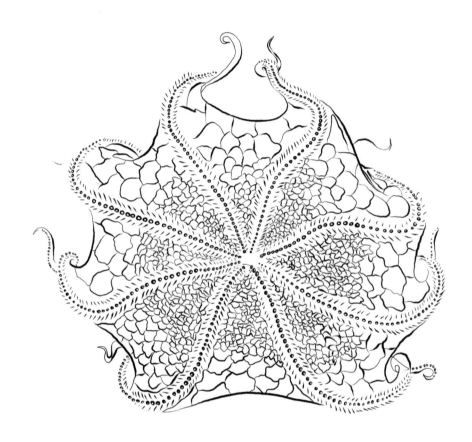

PROLOGUE

What would happen if we asked you to draw a map of the ocean? You'd probably highlight the shape—outlining the negative space between the continental shapes you know. You might color it blue because of the water. Maybe add some waves, for effect, or the sun rising over the eastern edge of the map. If we then told you the ocean is changing faster than it has ever changed before, yet none of those things are changing, you might reasonably ask: What do we mean?

Even though one-third of the people on this planet live within sixty miles of a coast, the ocean appears as a featureless expanse for most of us; it is an open blue palette separating the places where life happens. In reality, it is a mosaic as breathtakingly nuanced as the fields and cities and hills and valleys we know on land. And that mosaic is experiencing unprecedented change: coral reefs turning white, animals migrating to unusual places, ice sheets melting, and fish stocks that people have relied on for millennia dwindling. Scientists, Indigenous communities, conservationists, and fishers we have talked to over the last few years have told us of their desperation to communicate this change to the public and their frustration as the change continues to accelerate.

In 2017, we set out to document, take note, observe, and reflect. To listen. We could hear a change in the communities of people who were watching the ocean—an intensity shift, an urgency, a threshold being crossed. And so we did just that—we called people on the phone; met them on the beach; ate meals together; showed up to webinars, seminars, conference presentations; and, eventually, in the midst of a pandemic, conducted dozens of interviews online. We poured over manuscripts, books, and online videos, and emailed follow-up questions. People invited us into their living rooms, backyards, and neighborhood cobble beaches.

For this, we are especially grateful. In researching this book, we were entrusted with stories and histories of connection to the coast and ocean. We do the best we can to introduce you to these storytellers. We acknowledge that, in many cases, these stories are told by Indigenous knowledge holders, and in those cases, we are particularly careful to make sure the stories that are told here are told in the way that they wished.

In each of these discussions, we would start by introducing what we are trying to do with this book. The ocean is changing so fast, but people describe it as vast, open, blue, untouched, and untouchable. *Can you tell us how you are touched by the ocean?* The ocean is the main character in this story, we would say, but it needs a voice. *Can you tell us what you see changing?* There is bravery in being a witness to change. At its very heart, these are stories about the observers of the ocean, the people who know it best, and they want you to know what they see.

This book is organized by places; as such, it is a meander through ocean environments. But it is not meant to be comprehensive or textbook-like in its coverage. These places and stories are representative, not exhaustive. The challenges faced by people studying, learning, living, and loving these places—and the hope they carry for the future—are meant to be a starting point for exploring the changing ocean.

AT EVERY DEPTH

CHAPTER 1

THE TIDE POOL

Who has known the ocean? Neither you nor I, with our earth-bound senses, know the foam and surge of the tide that beats over the crab hiding under the seaweed of his tide-pool home; or the lilt of the long, slow swells of mid-ocean, where shoals of wandering fish prey and are preyed upon, and the dolphin breaks the waves to breathe the upper atmosphere.

—Rachel Carson, Undersea

We often hear how little people know about the ocean. The challenge with the ocean, of course, is that it is hard to wrap your arms around it: the expanse, the depth, the eternal blueness. Yet we should also not pretend that we don't know anything at all. What we know is, in fact, remarkable. The human exploration of the ocean over millennia, in every place where we could touch, observe, listen, poke, and prod, has led us to a considerable understanding of its nature. Now we face a new difficulty, one greater than even the physical or scale limitations we've come up against so many times before: the ocean is changing so fast. How do you understand change affecting two-thirds of the surface of our planet? How can we tell a story about the ocean and people and our collective future while simply sitting on the shore?

You can start, as our story will show, with a tide pool. The tide pool is an entire universe, rocks and sun and water, forests and shrubs and meadows, grazers and roamers and apex predators, all fully contained in a few feet of rock and reset with each change in tide. It captures, too, the ocean's extremes. It warms in the sun then chills rapidly once the ocean water returns. As animals breathe out carbon dioxide in an isolated pool, the chemistry changes dramatically, a preview of a changing future ocean worldwide. In a storm, a tide pool can be diluted by fresh water, and at high tide it flushes with salt. It seems an impossible place to live and yet it is, as John Steinbeck once wrote, "ferocious with life."[1]

Rocky tide pools make up a very small percentage of the Earth's surface. But they have been home to pioneering experiments in our understanding of the natural world and have shaped the minds of some of the brightest scientists, writers, and artists in history. Nearly one-quarter of the Greek philosopher Aristotle's surviving writing concerns his interest in marine biology, developed over decades spent peering into Mediterranean tide pools and consulting local fishers. The last two books that the English naturalist Charles Darwin published before announcing the idea of evolution in *On the Origin of Species* were deeply researched monographs about barnacles. Rachel Carson, who is often credited with kickstarting the American environmental movement with her 1962 book *Silent Spring*, spent the first decade of her professional life as a marine biologist for the U.S. Fish and Wildlife Service, observing tide pools as well as salt marshes and sandy beaches; the last book she wrote before *Silent Spring* was called *The Edge of the Sea*. The former National Oceanographic and Atmospheric Administration (NOAA) administrator Jane Lubchenco, appointed to the White House Office of Science and Technology Policy under President Joe Biden, launched her scientific career studying seaweed distribution and food web interactions in tide pools.[2]

Ideas about competition, cooperation, predation, and even animal behavior have originated in the tide pool, where scientists can watch a complete universe at work. If you've ever heard a wolf, lion, or shark called a "keystone species," you're appreciating a concept that scientists first gleaned and refined by watching sea stars hunt in the rocky mussel beds of coastal Washington.[3] Like peering at the stars in outer space to understand the rules of gravity, we can peer into a tide pool to map the invisible ecological forces that shape our entire world.

"Ecologists are often accused of having physics envy," Karina Nielsen, a marine ecologist and the director of Oregon Sea Grant, told us. "Because physicists model things and come out with simple models that have strong explanatory power. But that doesn't mean there aren't underlying principles that drive the [ecological] patterns that we see, and that they can't be understood."

As a species, we humans also have a longer relationship with the shoreline edge than with any other kind of ocean habitat. Scientists use tide pools as a model for ecosystems, but they serve just as well as an analogue for our connection to nature. The human ability to read nearshore change goes as far back, some scientists think, as the modern human mind.[4] Early in our history, as people were first starting to spread across Africa, the climate changed, and vast deserts formed in the interior.[5] Some early humans decamped to the rocky shoreline. At one cave site at Pinnacle Point in South Africa, paleoanthropologists have found shells, food, and tools dating to 164,000 years ago.[6] Based on calculations of past sea levels, the caves would have been a few miles inland at the time. To survive there and then, our species would have had to know or learn about tides and how to make a priority of a trip to the ocean. "I surmise that the people who lived at [Pinnacle Point] . . . scheduled their trips to the shore using a lunar calendar of sorts, just as modern coastal people have done for ages," the archaeologist Curtis Marean wrote in *Scientific American* in 2010. Although the field of human origins has many competing theories, Marean says that he believes the evidence from those caves showed that in a time of climate change and mortal peril for our species, the tide pools—the ocean—"saved humanity."[7]

Now, as the ocean changes more rapidly than it has in human history, the understanding of the tide pool we have built over millennia will matter. To measure change, you have to have a baseline to measure it against. Tide pools, one of the oceanic habitats we know best, may be one of the most valuable baselines that we have to understand the past and present global ocean.[8]

One of the world's most studied shorelines stretches 3,000 miles down the western edge of North America, from Alaska to Baja California. Winds swirling in the Gulf of Alaska cascade down the coast, stirring up the ocean and upwelling cold, nutrient-rich water that sloshes against the rocky edge of the continent. In many places along the rocky shoreline of the West Coast, there might be hundreds of visible species and thousands of individual animals in a few square yards. To take one particularly meaningful example: in the mid-1990s a small team of ecologists surveyed a narrow, 108-yard-long, one-yard-wide stretch in a tide pool area in central California. After two years of visits, they had counted 125,590 individual animals on the rocks, representing 135 different species. Even that was a limited total. The scientists admit they chose to ignore one species of clam, three abundant species of small snail, and all the various colony-forming tunicates and sponges, which were simply too difficult to count.[9]

This closely investigated tide pool in Pacific Grove, California, formed where a crown of granite pushes out northeast into Monterey Bay. The rock slabs lean at angles against one another like a giant tumbled set of children's blocks. In the deep fissures, cracks, and channels between the rocks, the blue water sloshes through. The incredible biodiversity in the tide pools feeds a host of larger animals that move through. Otters splash into the shallows; egrets and herons stand still in the water; gulls and oystercatchers perch on the mounds of seaweed that pile on the lower rocks. On the smooth, rounded tops of the higher rocks, cormorants spread their wings like decorative gargoyles overlooking a castle. The cormorants preside over a piece of shoreline with a variety of names today. Where the rock pushes out farthest into the bay, it forms Point Cabrillo, named for the European navigator who visited the California coast in 1542. The rocks and pools sit just inside the southern boundary of the Lovers Point–Julia Platt State Marine Reserve and just north of the Edward Ricketts State Marine Conservation Area. Directly above them, in a series of low, barracks-style buildings is Stanford University's Hopkins Marine Station.

People lived around and visited these tide pools for millennia before the arrival of Europeans. Today the Indigenous people of Monterey Bay go by a variety of names, including Rumsen, Rumsien, Ohlone, Costanoan, and Esselen. All but the Esselen once spoke Rumsen, a language based in a family shared with people to the north in the San Francisco Bay Area. Historians and linguists call these the Ohlone languages, and—although the people at the time likely would not have seen themselves as organized by their primary language—tend to separate groups today by the primary Ohlone dialect they spoke.[10] The Spanish called them all *costeños*, coast people.

On New Year's Day, 1603, it snowed in Monterey. The nearby Santa Lucia Mountains turned white. It reminded the Spanish navigator Sebastián Vizcaíno, who had anchored just off Point Cabrillo, of the Popocatepetl volcano in central Mexico, where his voyage had started.[11] Water left out overnight froze on the Spanish ships, and a crust of ice several inches thick formed on freshwater ponds. A small group of Spanish set off to explore the land around Monterey, hiking through the bitter cold as far as Carmel. They reported empty villages; everyone had moved inland to wait out the weather. Two days later, the Spanish were gone, sailing north up the California coast for Cape Mendocino, skirting the entrance to the San Francisco Bay that they did not yet know existed. It would be more than 150 years before they returned, this time to stay. In 1769, Spanish soldiers marched north from San Diego to look for the Monterey harbor Vizcaíno described. They had been sent to look for a place to establish an outpost against the English and the Russians, and Monterey seemed promising. By 1772, the Spanish had built a base in Monterey and a mission in nearby Carmel. The Spanish compelled the Indigenous people they found to relocate into the missions, centralizing people who had once spread out across the land according to the seasons. Spanish settlers banned the practice of burning oak woodlands, which had encouraged tree and grassland growth, and carved the land into ranches, introducing cattle herds and invasive pasture grasses.[12] In the summer and fall today, two centuries later, the hills of the Golden State turn gold from the sun shining on Spanish oats, fescue, and barley. For coastal Indigenous people, this part of history marked the beginning of a centuries-long resistance to colonial oppression and rule, and fighting to maintain their connection to the land and sea that had been honed over thousands of years. Even beyond the Mission era, "Native people creatively reaffirmed their connections to place and cultivated a sense of belonging within homelands increasingly becoming home to others," the Coast Miwok and University of California, Santa Cruz, archaeologist Tsim Schneider writes.[13]

Change in the ocean is not always so evident as change on land. The invasive weeds growing along the roadside are there for anyone to see, but only those with experienced eyes see the algae expanding in a tide pool or once-common shells becoming rare. Change, however, happens whether most people see it or not. Beneath its eternally blue surface, in the pools around its timeless granite, the ocean where Vizcaíno anchored in 1603 must have been shockingly different from the one you would see out the window looking at Monterey Bay today. And because people don't live in the water, it has been one of the great challenges in culture and in science for the witnesses who do see the ocean changing to convince the world of what they know.

In 1931, a young man named Willis G. Hewatt arrived in Monterey. Hewatt had just received a master's degree in biology from Texas Christian University in Fort Worth, Texas, and married his wife, Elizabeth. The day after the wedding, the couple drove west. Hewatt had decided to continue his academic work and pursue a PhD in the tide pools of Pacific Grove, California.[14]

He arrived in a town booming with cash, culture, and science. Monterey in the 1930s was home to a literary community headlined by John Steinbeck. When he wasn't writing, Steinbeck explored his love of marine biology in field classes in the tide pool behind the new Stanford University Hopkins Marine Station. The marine station, built on top of a fishing village after the eviction of the Chinese immigrants who had made a home there, had rapidly established itself as one of the nation's premier marine biological laboratories, a West Coast rival to the famed East Coast labs at Cold Spring Harbor Laboratory, New York, and Woods Hole Oceanographic Institute, Massachusetts. On the Monterey waterfront, fishers, boat operators, and factory workers thrived in the seemingly inexhaustible sardine fishery that gave name to Steinbeck's 1945 novel, *Cannery Row*.

Just south of Hopkins, near where the Monterey Bay Aquarium was built a half century later, a marine collector named Edward Ricketts operated a mail-order biology supply store called Pacific Biological Laboratories. Ricketts was working on a manuscript for a book called *Between Pacific Tides*, a first-of-its-kind guidebook to the tide pool that focused on life from shallow to deep, as a collector might find it, instead of arranged taxonomically. The manuscript was at first rejected; the director of the marine station at that time wrote that "the method of taking up the animals from the standpoint of station and exposure on the seashore seems at first sight very logical but from the practical standpoint it seems to me not particularly happy."[15]

But Hewatt, who read Ricketts's first draft in the Hopkins library, found the idea of a tide pool organized by zones intriguing. Hewatt decided he wanted to study the arrangement of animals in a tide pool for his dissertation. In 1931, he drove a series of four iron bolts into the Pacific Grove granite. For the next three years, he went out at low tide and counted and categorized all the tide pool inhabitants in one square yard around his bolts. The study site, stretching from a nearshore boulder below the marine station to a large mussel bed 324 feet out into the tide pools, became known as Hewatt's transect. Yet its creator couldn't imagine the significance his site would one day hold. Upon the conclusion of his work in 1935, he filed his dissertation with his committee, received approval of his work, and moved back home to Texas to become a professor of biology at Texas Christian University. Mats of purple and brown seaweed soon grew thick over the transect bolts. People knew Hewatt had been there but couldn't say exactly where. His dissertation was published in the *American Midland Naturalist* in 1937, and perhaps only few noticed.[16] After all, how could it matter?

On a clear winter day in 2019, the ecologist Jim Barry walked us from the Hopkins Marine Station along a dusty path to the edge of the bluff, past a pair of sea otter researchers standing over a spotting scope, and toward the tide pools. It's a scramble down a head-high drop from the bluffs to the beach, onto a thick layer of broken mussel shells that crunch near the water's edge. The rocks here date to around 80 million years ago, little broken-off pieces of volcanic deposits, bumping northwest over the eons in the fault zone between the Pacific and North American geologic plates. Worn and eroded, they nonetheless feel as eternal as the blue ocean around them. When he was a graduate student in San Diego, Barry would try to work in the crumbling and constantly shifting sandstone of the tide pools of La Jolla. "Granite," he said, appreciatively, noting the contrast. "It's here forever."

When Barry started his marine biology career in 1991 as a young scientist at the Monterey Bay Aquarium Research Institute, his office was right up the street from Hopkins, and he'd often come down to the lab to talk to the station's resident elder, Chuck Baxter, who'd been lecturing to Stanford University undergraduates since the 1970s. Barry was a scientist and a surfer who had been hooked by the ocean early as a kid growing up in Northern California. He had spent summers in the Antarctic as a research assistant and earned a PhD in oceanography from the Scripps Institution of Oceanography in San Diego. Baxter was a thinker and philosopher who knew the history of the place and everyone who'd come through it like he knew his own family. He'd worked up and down the West Coast, read deeply of Pacific history, and been one of the founding scientists at the Monterey Bay Aquarium. Once, Baxter mentioned to Barry that he felt like things had changed in the tide pools out the window. He looked out at the water pouring through the rocks, and he felt something about it looked like Southern California. They got to talking. Baxter had been thinking already about the Hewatt transect, and now he wanted to return to it. According to the predictions of the still-new science of global warming, you would expect marine creatures to move north toward colder water at the pole. Maybe Baxter's gut feeling that he saw Southern California out the window had a basis, one that Hewatt's historic descriptions of the tide pool might allow observers to unveil.

Every year, Stanford undergraduates would come down to the marine lab from Palo Alto and spend a quarter there doing research. Baxter saw a resurvey of Hewatt's transect as an ideal undergraduate project, and for several years running, he unsuccessfully pitched the idea of studying the transect to students. In 1993, Raphael Sagarin and Sarah Gilman took him up on it. Both were juniors, and unlike many of the premed students at the research station, they were interested mainly in ecology. Gilman, now a professor of marine biology at Claremont McKenna in Southern California, says that she had no particular tilt toward the ocean. It was her first marine biology class. She admits the career path she set off on then seems almost accidental now, that if Stanford's field course had been in the desert, she might these days be a desert ecologist. "But after that quarter," she says, "I never wanted to study anything else."

Baxter also recruited his friend Barry to join them. There was only one problem: no one could remember where exactly Hewatt's transect *was*. While Hewatt had described the transect carefully in his work, a clear line to sketch out the exact path of his study site was not included in his thesis. So Sagarin, Gilman, Baxter, and Barry plunged into the tide pool to look for Hewatt's bolts. Barry remembers being out at a low tide in the middle of the night, Coleman lanterns casting the granite in sharp relief, pushing piles of seaweed aside to sweep the exposed rock with metal detectors. Gilman, too, remembers the darkness, the quiet of the 4 a.m. low tide, the feeling of combing the seaweed and later trying to count all the species in her hand. "The number of different animals that you could pull out of a square yard out there—it was . . . magical," she said. "You have no idea there's that much diversity down there but if you work your fingers through every tiny bit of algae there's so much life."

They found the bolts eventually, and during low tides that year, Gilman and Sagarin would lay down a one-yard-square box at each bolt and count and identify everything within it. By the end of the spring, they had documented 58,000 individual animals. The next step was to compare what they had counted to the species and positioning Hewatt had described sixty years earlier.

In a 1995 article in *Science*, Barry, Baxter, and the just-graduated Gilman and Sagarin reported that the abundance of eight of nine warm-water southern species had increased at the site while five of eight cold-water northern species had decreased since 1935. The shift coincided with measurements showing that the water at the site had warmed over the decades as a result of climate change.[17] Although they had spent their time counting and sorting small marine creatures, the headline was much bigger. Ocean watchers didn't have to wait to document change—it was already happening. "I knew it was a big deal because suddenly we're being interviewed by CNN," Gilman said. "But in terms of it being cited by people who I look up to and admire as ecologists, it never occurred to me."

For centuries, many Western scientists and writers had described the ocean as so vast as to be beyond the human ability to disturb. We didn't need to wrap our arms and minds around the ocean, in some sense, because it was an ever-replenishing spring. Barry, Baxter, Gilman, and Sagarin arrived with evidence at a turning point in the scientific relationship to the ocean. Other scientists had just begun to grapple with the implications of global warming and, because the tide pool resurvey had connected the present to the past, it landed with particular heft. "During that period we really started thinking about how marine systems are changing," Barry said, "and the future is not one of stasis and stability."

Nearly three decades later, the human pressure on the ocean has only become more and more apparent. Researchers inspired by the study, who started watching their own research sites more closely after 1995, have now had time to record slow-moving change of their own. Scientists are trained to be observers, to record and document and measure where we've come from. In writing, researching, and listening to these stories, we have begun to see their work as an effort to preserve stories—to drive anchors in the granite to provide future generations with a map of a world we once knew.

In the 1830s, Mexico and California became independent from Spain, and the missions were secularized. The Ohlone of Monterey Bay began to make their own way in a world transformed. By the 1840s, Americans began to arrive and claim land of their own or take it by force.

In 1846, an Ohlone woman named Isabel Meadows was born in Carmel. Her mother was a Rumsen speaker who had been born in the mission, her father an English whaler. She grew up in the valley with her mother, grandmother, and great-grandmother, learning a combination of old ways and new. Her life spanned the Gold Rush, American statehood for California, the extirpation of the grizzly bear from the new state, the first calculation of the greenhouse gas effect, and World War I. Hewatt's first view of the Pacific Grove tide pools nearly coincided with Isabel Meadows's last look at them. In 1934, when she was in her eighties, she traveled to Washington, DC, to work with a Smithsonian ethnographer named John Peabody Harrington. Harrington spent the last years of his life trying to interview Native Californian survivors to document vanishing Indigenous languages. Meadows, the last fluent speaker of Rumsen, spent the last five years of hers telling him everything she could remember. In the same years that Hewatt had spent recording one kind of message to the future, Meadows and several other Ohlone elders gifted future generations the information they'd need to revive their culture.

"Isabel's mind contained information about this specific California Indian community never before recorded, gleaned from her own long life, her mother's, her grandmother's, and many significant elders," writes the Ohlone Costanoan Esselen Nation author and historian Deborah Miranda.[18] "Isabel has a clear purpose in depositing these stories with Harrington: to preserve information from Ancestors in ways she knew would provide necessary information for future generations."

Handwritten in a mix of English, Spanish, and various Ohlone languages, Harrington's notes went into loosely organized boxes in the Smithsonian and eventually onto microfilm shared with a small number of university libraries. Meadows died in 1939, and Harrington in 1961. By the time a Rumsen woman named Linda Yamane went looking for the Harrington notes in San José in the 1980s, only a handful of people remembered the history or the record of it.

Yamane did not grow up with the songs, stories, language, or environment of her Rumsen Ohlone heritage. She was born in 1949 in San José, a few dozen miles north of Monterey, where her father's mother's family had survived the Mission period and American colonization. Her grandmother told her stories from the turn of the nineteenth century, but she didn't recall much about Rumsen culture.

As she grew older, Yamane's curiosity about her own history increased. She visited the missions, where the birth and death records of her ancestors had been recorded, but "always came away feeling empty and sad," she wrote in an essay called "Lost and Found: Ohlone Culture Comes Home."[19] Then, in 1985, someone told her that the San José State University library held a treasure trove of mostly unexamined Ohlone information. It was the microfilm archive of Harrington's work: vocabulary lists, field notes, and the stories Meadows and a handful of her contemporaries had told the ethnographer.

It was surprisingly difficult for Yamane just to get an appointment to see the microfilm. Librarians gave her conflicting information. When she arrived for the first time, the person who knew about the microfilm had gone, and the librarian on duty hadn't heard of it. She told Yamane to come back another day, but Yamane wasn't ready to accept defeat. "I just couldn't let go," she told us. "I was so close and yet somehow they were out of my reach. I was being very stubborn within myself and just couldn't leave. I began walking up and down the aisles and thinking, 'They've got to be here somewhere.'"

Luckily, the librarian returned and guided Yamane to the microfilm. And through the 1980s and early 1990s, as the Hopkins scientists brushed through seaweed in the darkness, looking for the bolts unvisited since Hewatt left them, Yamane pushed through the darkness of the library stacks in a search for Rumsen knowledge rarely accessed since Meadows dictated it in the 1930s. Later, Yamane wrote about how she felt shy saying her ancestors' words out loud. She would practice them late at night, sitting awake in bed while her family slept. From basic words she moved to songs, which had been recorded on wax cylinders in Monterey in 1902 and are now archived in the Phoebe Hearst Museum of the University of California, Berkeley. After studying museum collections and talking to basket makers from other Northern California tribes, she became the first Ohlone to weave an Ohlone-style basket in about 150 years. Now a renowned master basket weaver, she has written that one of the most important motivations for making traditional baskets is that "most of our old baskets are gone, but now people can look at one of my baskets, hold it, see the intricacies, and know that our ancestors were resourceful, intelligent, artful people."[20] Interested in the canoes in which the Rumsen navigated their watery world, in the late 1980s she joined a group to build one of the first tule canoes to ply the San Francisco Bay since the 1800s. Since then, she has built more than thirty new tule canoes, including several for a summer program at the Monterey Bay Aquarium. For several years, the canoes were paddled through the Great Tide Pool, an artificial rocky pool fronting Monterey Bay a few hundred feet south of Hopkins Marine Station.[21]

In the mid-1990s, Yamane also became interested in Ohlone jewelry, which relied heavily on coastal resources. From historic drawings and archaeological materials, she inferred that the Rumsen wore ear ornaments and necklaces using abalone and *Olivella* shells. She also needed to make hundreds of tiny sequin-like *Olivella* beads for some of her finest baskets. But this is the challenge when everything has changed, when no one alive knows how to make the thing you want to make. Where do you look? How do you even start? What did your ancestors know that has been lost? When we asked Yamane what Rumsen speakers like Meadows had said about the ocean, she told us she recalled mainly hints that spoke to a deeper connection. Meadows's family members would gather salt and seaweed from Carmel, south of Monterey Bay. Meadows's great-grandmother's first husband had been an abalone diver and had drowned off Point Lobos. In the post-Mission period, a man would ride his horse up from Pebble Beach with fish for the Meadows family and speak to Isabel's mother in their language and trade fish for food from the farm.

Yamane started to reforge the connection so she could make jewelry. She went to the beaches and explored. She found two sites for *Olivella*, one in Monterey, one in Carmel. To her surprise, the shells at each site were quite different. The ones in Monterey were smaller and whiter. The ones in Carmel were larger, sometimes "humongous," she told us. They tended to be mostly brown or gray, less useful for traditional purposes that prized white shells. Had it always been this way? Did her ancestors know the variation in *Olivella* shells up and down the coast? Did they have favorite collecting spots where they would find the shells that were just right?

She learned to heat-treat the shells in hot sand. It further whitens them, Yamane told us, but also changes the texture, making the shells more chalky and less brittle. "*Olivellas* will fracture along their vertical growth lines," Yamane said. "You lose a lot of material when you're trying to make beads, because so many of them fracture. It was through a lot of trial and error that I learned, through practice and reverse engineering."

She also worked with abalone, the marine snail whose iridescent rainbow shell humans around the globe have prized for tens of thousands of years. Northern Californians today line their gardens and fences with them. Yamane knew she could find broken pieces of shell here and there, but acquiring them in the supply she needed for jewelry was a challenge. By the time she was looking seriously, the Southern California red abalone fishery had nearly collapsed, and Northern California permitted only recreational free diving and a total limit of three red abalone per season. By 2017, even that fishery had to close.

Yamane had a friend in Berkeley who would go to the city's big flea market and see abalone shells for sale. The friend would buy them all and send them to Yamane in bulk. Later, she met some abalone divers who emailed around, asking fellow divers if they had discarded abalone shells in their yards. Several more cardboard boxes, stuffed with shells, arrived. "At the time I didn't want to be greedy," Yamane remembered thinking. "But since they were offering, I thought I should accept because my community might really need them some day."

Yamane established herself as a sort of abalone bank just as the abalone were disappearing in the wild. "At that time I wasn't thinking about climate change but just the fact that so many natural resources become less and less and less over time," she said.[22] "I'm seventy-three, and that's enough years to realize how much things have changed." Barry, Sagarin, Gilman, and Baxter don't mention abalone in their 1995 paper, and abalone merits only a brief mention in a follow-up study done by Sagarin and colleagues in 1999. We saw only a handful in the tide pool on our walk with Barry, all small. But Yamane's ideas about how to respond to change and loss resonated. Over three decades, Yamane has continued to reclaim those stories from the past and make them part of the present. So a few hours after visiting the tide pool, where Barry and his colleagues had received one kind of message from the past, we drove across town and had dinner with Yamane. At a small sushi place near her house, she talked to us about the work it takes to find and revive a lost connection. She wore, as she often does, her abalone earrings.

Yamane expressed her gratitude for the Ohlone who possessed the courage to keep watching and the foresight to record and document their world. "Obviously it takes a lot of work on our part," she later told us. "But the first necessity is that the information is there in some form, that it was preserved." You can restore something vital to the world, Yamane said, by taking that documented information and turning it into something people can see and touch. That's how she characterizes her own work and career: half historian, half artist, trying to create new beauty out of past knowledge. "The information itself is tangible but I want to transform that into something three-dimensionally tangible," she said. "Something you can hold. That, to me, is when it comes to life."

Even in a habitat as studied and visited as tide pools, it can be hard to document change. Baxter had a feeling that the Pacific Grove tide pool had changed because he'd been there so long. Yamane had a feeling that the coast had changed because her ancestors left hints about the relationship they'd once had. And because Hewatt had left behind a record from long ago, it was possible in one spot to quantify the change. The Pacific Grove tide pool is well known to scientists as a result. But there are thousands of miles of rocky shorelines around the world, thousands of local observers who spend their lives at the coast and feel like things are changing, but they don't always have the quantitative evidence to show it. Species that move on or die off don't always leave obvious gaps behind. New life moves in. "If I don't have a personal experience with it, it doesn't look like something's wrong," says University of California, Irvine, marine ecologist Cascade Sorte. "It's not like you're out there, and there's a bunch of dead mussel shells on the rocks, the legacy of past mussel beds. It's been taken over by other species, whether that's good or bad."

James Carlton, an emeritus professor of marine biology at Williams College in Massachusetts, has documented cases from around the world in which species that once appeared to scientists to be fundamental to a particular place are in fact relatively recent arrivals.[23] "I have often asked my students to imagine what the shore they are standing on with me—whether it's in New England, or the Pacific Northwest, or California, or Louisiana—may have looked like only 500 years ago," Carlton wrote to us in an email. "Despite all that we think we know, my guess is that we would gasp if we could be 'beamed back.'"

Sorte tried to take on this challenge off the coast of Boston in the early 2010s. Longtime locals told Sorte and her collaborators that the coastline's characteristic blue mussel beds had declined over the years. The question, as always, was, How to go back in time to look for a quantitative comparison to the present day? Sorte and her colleagues found seven studies had been done on the Gulf of Maine in the 1970s, covering 250 miles of coastline. In 2013 and 2014, they returned to the shoreline to count mussels and complete the comparison. Across each site, mussels had declined dramatically. "Over 40 years, mussels have gone from being a defining species of intertidal habitats in the Gulf of Maine to being a minor contributor to compositional patterns and a spatial subdominant," Sorte and colleagues wrote in 2016.[24] As the Gulf of Maine warms, it seems the mussels too are moving northward. The southern edge of their range has already contracted—blue mussels once lived commonly as far south as North Carolina and Delaware, but those populations are mostly locally extinct. There's evidence too, Sorte wrote, that blue mussels are expanding their presence in the Arctic. As they move, shorelines are rewriting themselves within a human generation—and still, it's not a change visible to most people.

"The people observing it for a long time have a longer history, and so they've seen it change, perhaps gradually," Sorte told us. "It's a 'you don't see yourself growing day to day' type deal. They have a sense of the immensity of the change, more even than the researchers who are trying to come up with this picture in our heads of what it used to look like." What looks normal today wasn't many years ago, and likely won't be in the future. Yet the human tendency to feel that whatever world we see is the normal one and to measure change in any direction from that often conceals rapid, dramatic ecological decline, a process identified as a "shifting baseline" in the fisheries literature.[25]

At the California Academy of Sciences in San Francisco, the marine scientists Rebecca Johnson and Alison Young had another way to approach the shifting baselines challenge. Millions of people visit the California coast every year, and the vast majority of them visit with an internet-connected smartphone and its built-in camera. If some fraction of those millions could be persuaded to take a few pictures of what they see at the coast and add those pictures to a global database, it might both build a baseline for the California coast and create more ocean observers.

For a few weeks every summer, the scientists run Snapshot Cal Coast, a coordinated effort to get people out into the intertidal zone, on the sandy beaches, and on the bluffs and marshes of California to observe and take note.[26] Participants log their ecological observations in iNaturalist, a global community science app that now holds more than 160 million observations from more than 2.8 million people. The app has a large community of expert identifiers who help sort and identify all the photos and uses artificial intelligence, trained on the millions of verified observations in its database, to offer rapid IDs. Thus, Snapshot Cal Coast builds on two incredibly powerful platforms that have emerged recently in science: the power of big data sets, making it possible for scientists to understand trends through time and space, and the power of community scientists. The data collected by coastal observers creates a snapshot in time of what species are present along the coast—hundreds of thousands of photos from a specific time and place that can be used by scientists and managers to understand biodiversity and the pace of change. The first Snapshot Cal Coast launched in 2016 with 7,374 observations in a two-week span in June. In 2020, as people sought outdoor refuge from the COVID-19 pandemic, Johnson and Young chose to make the Snapshot a nearly six-month event, netting almost 200,000 observations.

Johnson and Young also create an annual "most wanted" list of species that are particularly hard to find, that have limited data available, or that we know are moving because of climate change, making it more important to track their every step. From the beginning, these lists have featured tide pool species making the poleward march up the state—like the California spiny lobster, a once exclusively Southern California species now regularly found north of San Francisco Bay, and the bubblegum-pink Hopkins rose nudibranchs, once like a celebrity sighting for a Northern California tide pool observer, that now seem to live there. Recent "most wanted" lists have included anemones like *Anthopleura elegantissima*, which were highlighted by Sagarin, Barry, Gilman, and Baxter in their *Science* paper and in a follow-up paper in *Ecological Monographs* as a species sensitive to temperature-driven range shifts.[27]

———

Early in his career, Barry spent much of his time in Pacific Grove, on the rocky shore looking into the tide pools. But when we met, it had been almost twenty-five years since he worked regularly in this exact tide pool, and he spent half an hour slowly reacquainting himself with what he saw. He remembered the giant green anemones, *Anthopleura xanthogrammica* and told us they might live to be one hundred years old, staying relatively close to home, probably within one pool, for their entire lives. Over the last few decades, a striking striped species called the sunburst anemone, *Anthopleura sola*, has been taking over more space in the tide pools. Barry felt like there were more of the giant pink barnacles called *Tetraclita* than there had once been. He commented on the color of the purple brown algae mats clumped over the rock. Change in a tide pool may be subtle, aesthetic really, but noticeable to the trained eye.

We looked into a crevice in one of the rocks, a dark slash extending several feet down to the water. It was lined with sea stars and more pink barnacles. Below the barnacles, we saw the characteristic curving, dotted shells of a handful of abalone. Across the channel, mussel beds rose over the water, surf pouring through gaps between them. A gull perched on one bed and pecked at the candy-like remains of a gumboot chiton.

Barry's career turned long ago from tide pool studies to investigating the slow-moving deep sea. He could take the lessons and even the methods from what he'd done in the intertidal zone, he realized, and apply it to the darkest depths of the ocean. There was no Hewatt transect for the deep sea, so one would have to be created. The tide pool would be the scaffolding for so many other studies, and as we visited the tide pools now, it reminded him of the transient nature of the seashore. "When I come here," Barry said, "I see what's changing."

Gilman told us, when we talked to her twenty years later, that the student project had become a career-defining inspiration. One of the critiques of their paper had been how local it was, so as a graduate student at the University of California, Davis, she decided to look at the effects of temperature on a single species, the rough limpet, caught in the transect on a northward march. This carried her into a postdoctoral position at the University of South Carolina, a second postdoctoral fellowship at Friday Harbor Laboratory in Washington, and a lifelong interest in temperature and heat stress and how to predict the ways they affect marine animals. In 2010, she wrote a critically important paper for the field of community ecology, "A Framework for Community Interactions Under Climate Change."[28] As a professor at Claremont McKenna, she teaches science courses to undergraduates, where sometimes she assigns the Barry et al. 1995 paper—although she told us that it sometimes takes students a while to figure out that one of the authors is standing there in front of them.

Even in the 1990s, Gilman told us, people didn't necessarily think of species moving because of warming as a threat. But twenty-five years of science have shifted her perspective. She's learned more, she said, about the sometimes fragile nature of ocean life. She watched a marine disease outbreak claim 90 percent of sea stars on the West Coast in three years.[29] She had a study site go from being a mussel bed to an anemone bed and, because there wasn't long-term monitoring there, had no idea whether that was normal or not. She began teaching an undergraduate class on how science is communicated to the public. When she started, she said, the room was often evenly split between students confused by or dismissive of climate change and those who accepted it. Now, she says, the majority of her undergraduates accept it as reality.

When Barry, Sagarin, Gilman, and Baxter's paper came out in 1995, the Pacific Grove tide pool became a message from the future and a rallying point for scientists and ocean conservationists. In 1998, President Bill Clinton and Vice President Al Gore asked to see the tide pool while they were in Monterey to announce a series of ocean conservation measures. In pictures from the tour, you can see Sagarin in his jeans, shirtsleeves rolled up to the elbows, brown hair curly and unkempt, leaning over a small pool opposite Clinton in the slacks, dress shoes, and golf shirt that pass for presidential tide pool attire.[30]

Sagarin would later go to work for the Department of Defense, translating what he had learned in this tide pool about risk, organization, and adaptability into post–September 11 antiterrorism measures. He wrote a book, *Learning from* the Octopus: How Secrets from Nature Can Help Us Fight Terrorist Attacks, Natural Disasters, and Disease. He and Baxter spent a year retracing Steinbeck and Rickett's journey through the Sea of Cortez.[31] Sagarin moved from Washington, DC, to Arizona to join scientists on the Biosphere 2 project, where he directed a marine program responsible for a fully operational 690,000-gallon "ocean" that he hoped to turn into a working model of the tide pools in the Sea of Cortez. And every year he would come back to Monterey to resample the Hewatt transect.

In 2015, a drunk driver hit and killed Sagarin as he rode his bike outside Biosphere 2. Scientists, journalists, friends, and fans around the world mourned the loss of a ceaselessly energetic polymath who had already changed the world for the better and seemed to have so much more to do.[32] "Rafe wanted to put things together," Barry said. "He wanted to understand nature. He went from, 'What animal is that?' to 'How does this play out?' to 'What hypotheses can we enact that will explain what we see?' And that way of thinking becomes 'How is all of Earth happening right here?'"

Sagarin had always kept his datasheets, his square-yard quadrat, and his tools for working the transect in the office of a Stanford colleague, the ocean scientist Fiorenza Micheli. After his death, Micheli took over studies of the transect. The codirector of Stanford's Center for Ocean Solutions, Micheli has worked around the world, with research projects on the U.S. West Coast, the Mediterranean, the Pacific Line Islands, and Baja California. She has worked on major questions and issues facing the ocean, including ocean acidification, hypoxia, fisheries management, and deep-sea hydrothermal vent ecology. And she has seen the ocean change. Micheli worked with Sagarin observing intertidal animals in Baja California. After a series of extreme heat waves and toxic algal blooms, the rich biodiversity she had once seen simply disappeared. Luxuriant kelps forests crammed with sea life turned to bare rock. She watched the same process happen in Northern California a few years later as heat, algal blooms, and marine disease caused kelp forests to shrink. When we talked to her in 2022, she had just returned from diving in Italy, where extreme heat had caused coral, sponges, and seagrass meadows to wither. Even the Hewatt transect, she told us, seemed to change over the years. Mussels had taken over the site, possibly because a disease outbreak had killed the sea stars that prey on them. Biodiversity along the transect had dropped. In a 2020 article reporting on what they'd seen since Sagarin's death, Micheli and a number of colleagues—including Baxter—wrote that the ability to document those changes showed the value of field stations.[33]

A scientific presence along the coast allows people to see and record change at a time when it's more critical than ever. The work of continuing to document changes on the Hewatt transect has become a community endeavor, with scientists, students, and volunteers working together to study the changing tide pool. "Our plan is to maintain it forever if we can," Micheli told us.

People tend to think of the ocean as a static place. Perhaps that's a product of the history of Western scientific oceanography, which began in the late 1800s with measurements of global average temperatures—the kind of world-spanning measurements that emphasize stability and gloss over variability. Or perhaps it's because so many of us have seen images of Earth from space, "showing the world's ocean as a uniform deep blue," one group of oceanographers wrote in 2018.[34] Yet the more we measure it, the finer our observations become and the more scientists witness extraordinary variability. The wider ocean is more like a tide pool than we've given it credit for.

Marine ecologists get to know their field areas, their tide pools, or their estuaries much like we know our neighborhoods. The scientists who immerse themselves in understanding the sea come to resemble the coastal communities who forage and depend on resources from the sea—they know the tides, the feeling of a big swell approaching, the shift in the colors on the rocks or in the water. They know, like Linda Yamane knows and like her ancestors knew, the subtle shifts in species from tide pool to tide pool and from beach to beach, the tiny quiverings of individuality and place in the seemingly infinite and uniform blue that covers two-thirds of the planet. In the paper describing her Rumsen revival project, Yamane wrote, "It wasn't an academic decision to learn our language, or to learn our songs, or our stories or basketry. It evolved. It was a calling, something I could not ignore or stop. This work is done because it has to be done. When you follow your heart, it takes you where you need to go."[35]

Many scientists also hold out hope that by witnessing and documenting environmental change, they can provide the tools for someone, someday, to undo it. Karina Nielsen told us that when students ask her about intertidal systems she has studied and observed for decades, she isn't sure what to say. The relationships between the ecosystem and its inhabitants have become hard to predict from year to year.

From 2014 to 2016, a remarkable marine heat wave swept across the California coast, bringing water temperatures three to seven degrees Fahrenheit warmer than normal. These warm waters, along with temporary reversals in typical current direction that allowed currents to flow from the south to the north, provided an ocean highway for southern species to march northward.[36] In July 2014, the scientists Jackie Sones and Eric Sanford at the University of California, Davis, Bodega Marine Laboratory—about a three-hour drive north of Pacific Grove—began observing unusual species on their regular beach walks. Sones began documenting these southern wanderers in her popular blog, *The Natural History of Bodega Head*.[37] What started as sporadic blog posts through the summer and fall of 2014—a purple striped jellyfish, blue buoy barnacles, tropical snails, a clear floating blob with tentacles called a siphonophore—began to come together into a running record of change.

Sanford and Sones's expert observations grew into more focused surveys. They used museum collections, including the expansive and well-documented collections at the California Academy of Sciences, to determine where species had been previously recorded. To fit the puzzle together, they worked with colleagues to add in measured sea-surface temperature data and radar-based wind information to show the flow of warm water along the shore. In March 2019, Sanford, Sones, and colleagues published a paper in *Scientific Reports* documenting changes in thirty-seven species as the heat wave persisted.[38] Sanford was one of the many people who worked with Sagarin at Hopkins. The intertidal species he and Sones observed during the marine heat wave in Northern California include three of the eight species that Barry, Sagarin, Gilman, and Baxter documented as northward-moving species in Monterey in 1993. They were some of the same species Barry had led us to in the Pacific Grove tide pool. While many of the range shifts caused by the marine heat wave were temporary, in a few cases, southern species were able to take hold. The sunburst anemone, with its radiating lines on its central disk moving toward its tentacles, is among a handful of species that have made a permanent home in Northern California, long after the effects of the heat wave dissipated. Even after temperatures cooled, at least five of the thirty-seven species have "maintained small but stable northern populations," the scientists Erica Nielsen and Sam Walkes wrote in 2021.[39]

The heat wave of 2014–2016 that restructured California ecosystems and accelerated the northward march of species was remarkable at the time, but we may look back on it as the crossing of a threshold into more common heat-wave events. Some observations indicated that temperatures took years to return back to "normal"; in fact, heat-wave conditions were again declared in Southern California in 2018, with ocean temperatures breaking century-old records.[40]

In 1956, Rachel Carson described the zones of the ocean as separated by a series of invisible gates.[41] Everything lived where it lived for ecological reasons, and those reasons turned the seemingly uniform blue into a mosaic marked by innumerable boundaries. The rules that scientists documented in the tide pool led to a better understanding of where those boundaries existed in the global oceans. The lesson of the past decade is that the gates have swung wide open. Change in the tide pool has come to stay. If it has happened in the tide pool, where we can see and measure it, what has happened out in the blue beyond the shoreline?

CHAPTER 2
THE REEF

There couldn't be a more important question in the history of our species than whether or not we destroy the planet's biosphere.
—Ove Hoegh-Guldberg, quoted in Coral Whisperers, *by Irus Braverman*

In January 1995, while Rafe Sagarin and Sarah Gilman were still counting animals in tide pools in Pacific Grove, California, research divers on the island of Saba, at the northern edge of the Caribbean's Lesser Antilles, noticed strange lesions and dying, broken branches on a kind of coral called a purple sea fan. Saba is a remote five-square-mile island consisting almost entirely of an extinct volcano called Mount Scenery. It has a total population of less than 2,000, and the reefs that surround it have been protected since the 1980s, making it an international scuba diving destination. Whatever disease was killing Saba's corals, the Dutch researchers who first studied it later wrote, it appeared to be already widespread around the marine park. A monitoring group called Caribbean Coastal Marine Productivity (CARICOMP) had scientists stationed at forty labs in twenty-one different countries, and sent divers out to look closely at their own reefs.[1] By late summer, sick sea fans had turned up in Trinidad, Panama, Puerto Rico, the Dominican Republic, Jamaica, the Virgin Islands, and the Florida Keys.[2] However and wherever it started, the disease had already spread across 1,500 miles of ocean. Drew Harvell, a marine biologist at Cornell University who specialized at the time in coral immune defenses, recalls a colleague warning her that if she wanted to collect samples, she'd better hurry. The sea fans were disappearing quickly.

The outbreak fit with Harvell's growing interest in the defenses that marine creatures employ—spines, poisonous chemicals, and stinging tentacles—and how those defenses could ramp up in response to a predator or pathogen. After growing up and attending college in Alberta, Canada, Harvell had moved to the University of Washington and Pacific Coast tide pools, where she studied a bryozoan that could build a phalanx of spines to ward off the nudibranchs and sea slugs that might try to eat it.[3] From spines she moved to chemical warfare, this time between sponges and corals in the Caribbean. Harvell wondered whether the defenses the corals created might also work against other environmental threats like disease. There was a practical implication: Harvell thought the chemicals she had identified in marine life might also be able to attack diseases that harm people. She spent a decade at Cornell University trying to isolate them and then set them up against bacteria in the lab. As she advanced the science, she started to think more about the interaction between chemicals and the microbiome of the coral itself. In the mid-1990s, as the Caribbean sea fans died, Harvell saw an opportunity to test her ideas further.[4]

By late 1996, a University of South Carolina Aiken microbiologist named Garriet Smith had tentatively identified the pathogen affecting the sea fans as a fungus in the genus *Aspergillus*.[5] Researchers later narrowed it down to the species *Aspergillus sydowii*, which lives more commonly in terrestrial soils.[6] A closely related species of *Aspergillus* causes lung disease in people, with the severity of the disease closely related to the health and genetics of the person who catches it. For the immunocompromised, aspergillosis is a serious and sometimes fatal threat.[7]

Smith told Harvell that there was an active outbreak in sea fans around the island of San Salvador, Bahamas, where Harvell had based her field research for the past decade. Smith regularly worked there, too, and was interested in a joint project looking at the disease beyond the confines of the lab. Harvell and a postdoctoral researcher, Kiho Kim, now a coral ecologist and professor at American University, arrived in the Bahamas in January 1996. Harvell and Kim landed at the airstrip in San Salvador, drove out to the research station, then donned their scuba gear to inspect the site. They first swam over a healthy area, "like a garden," Harvell later wrote, and toward deeper water. On the outer edge of the reef, they found what they were looking for. "We swam over to a section of reef that looked like a sea fan graveyard," Harvell wrote in her book *Ocean Outbreak*. "Fan after fan had been stripped of all living tissue and reduced to skeleton—stumps with bare branches sticking up eerily. These fans had died so recently that their skeletons had yet to be grown over by algae."[8]

As the tide pool researchers had done in Pacific Grove, Harvell and Kim set up transects in San Salvador, the Florida Keys, and a handful of other islands in the Eastern Caribbean to track the health of purple sea fans. Over the next ten years, Harvell, Smith, and Kim documented the peak and then decline of the *Aspergillus* fungus, and the decimation and then sporadic slow recovery of sea fans in the Bahamas, Florida Keys, and Yucatan Peninsula.[9] But the pathogen would ultimately turn into a secondary story. For Harvell and many other ecologists working on Caribbean corals, the parallels to human disease turned out to be key: with corals as with humans, the less stressed the immune system before the disease strikes, the better the outcome.

Harvell and Kim's transects, like Willis G. Hewatt's transects on the Pacific Coast, were fortuitously placed to catch the most intense stress test on coral reefs that—to that point—people had witnessed. In 1997, the oceans started warming into one of the strongest El Niño events ever measured. El Niño cycles start with unusually warm water along the Equator in the tropical Pacific, with the heat slowly rippling out through the world oceans. For marine scientists who had just read Jim Barry's article in *Science*, the signals of the heat wave to come portended an unprecedented reshuffling of the marine environment. For coral reef biologists, it would come to mean something else entirely. Many now point to that El Niño in 1997 as the beginning of the new regime for the ocean.

Harvell and Kim were swimming a transect in the Florida Keys one day when they noticed the corals turning white. "A spectacular change happened when a bright purple soft coral, *Briareum asbestinum*, suddenly turned overnight, from dark purple to a whiter shade of pale," Harvell writes in *Ocean Outbreak*. "What happened over the next week was unexpected. The skin on the bleached corals began to fall off in necrotic patches, followed by mortality. All told, an average of 68 percent of the bleached colonies died. We immediately dropped all our other underwater work to focus on finding out what was killing them."[10]

Many of the coral reefs in the Caribbean are considered fringing reefs, shallow areas of coral that run parallel to the shoreline a swimmable distance from the beach. A separate system of connected reefs, the Meso-American Reef, extends along the Central American coastline from the Yucatan Peninsula in Mexico to Honduras to form the second largest barrier reef system in the world. While reef biodiversity in the Caribbean is lower than in other coral systems—for example, in Indonesia, or the Australian Great Barrier Reef—the Caribbean is home to the highest biodiversity among Atlantic coral reefs. It is home to numerous, fast-growing elkhorn and staghorn corals called Acroporids.

Or at least it was. The Caribbean today is also one of the most famously damaged large area of coral on the planet. Pollution, overfishing, tourism, and sedimentation have reduced coral cover by at least 50 percent since the late 1970s. Although the sea fan outbreak in the 1990s was unusually widespread, disease outbreaks were common at the time and have become even more so. Acroporid corals dominated the Caribbean for 250,000 years, surviving and thriving through all manner of sea level rise and warming events. They are now nearly absent in shallow areas, leading the sea into what a team of international researchers in 2020 labeled an "unnatural state."[11]

Rebecca Albright, a researcher at the California Academy of Sciences in San Francisco, spent the first years of her career diving the coral reefs of the Florida Keys. She told us a story about an Australian coral researcher visiting Florida, snorkeling over the reef they worked on, and coming back to the boat to say, "Where's the reef?" When Albright later started to work at the Academy of Sciences and dove at Indo-Pacific sites in Pohnpei and the Philippines, she saw ecological relationships that until then she'd only read about.

Although coral reefs can be found in tropical oceans worldwide, the South Pacific remains the place for textbook diversity. More than 600 species of coral live in the so-called Coral Triangle, the roughly triangular region between the Philippines, Indonesia, and the Solomon Islands that is the most species-rich area in the entirety of the world's oceans.[12] Coral reefs extend south from the Coral Triangle into the Great Barrier Reef, the largest living structure on Earth, arcing south along the east coast of Australia. Made up of nearly 3,000 individual reefs and 600 islands, the 133,000-square-mile Great Barrier Reef extends from the Torres Strait between Australia and Papua New Guinea to the cooler waters north of Australia's Sunshine Coast. The individual reefs range in size and shape, depending on how they formed and where they rest in the larger system. Albright told us that one of her "favorite places on Earth" is One Tree Island, a ten-acre "yolk" of coral rubble nestled in an egg-shaped reef sixty-five miles offshore, part of a chain of islands in the southernmost section of the Great Barrier Reef. The islands formed as part of a vast, exposed coastal plain during the last glacial period, when sea levels were lower. As the water rose and swamped the coastal plain, reefs began to form, sand gathered around the reef edges, and coral rubble formed small islands. Research suggests that this growth of reefs happened in multiple phases, when conditions were right. When temperatures were too warm or sea level rise too fast, the reefs were stressed and didn't grow as much.[13]

Aboriginal Australians and perhaps Torres Strait Islanders—recognized by the Australian government as the Great Barrier Reef's Traditional Owners—navigated the area around One Tree Island in canoes and fished the southern reefs for turtles and dugongs. After colonization, Australians considered mining the southern reef for its oil. But when the Australian government created the Great Barrier Reef Marine Park in 1975, it set aside One Tree Island for scientific research. A small research station, managed by the University of Sydney, allows up to twenty scientists at a time to live and work in one of the most protected, remote, spectacular coral reefs on Earth. "It is one of the most unique places I've ever been on the planet," Albright says.

To get to One Tree Island, most researchers take a boat from the mainland across deep blue and often choppy open ocean. After a few hours, the lagoon and island loom turquoise and green over the blue. The reef, a three-mile by two-mile oval that fully encircles the island, can only be crossed at high tide, so they sometimes have to wait just beyond the foaming water, staving off seasickness, Albright says, by locking eyes on the Pandanus trees that give the island its name. Riding directly over the reef, you can look down at the coral and fish just a few feet below, and then the bottom drops away again into the lagoon, midnight blue where the water runs deep over sand, speckled with colors over shallower sandy areas and coral outcroppings. "One Tree is stunning," says Maria Byrne, a marine ecologist at the University of Sydney who directed the small research station on the island from 2003 to 2018. "It's beautiful. When you see One Tree, you realize you have this extraordinary diversity of species."

Sea eagles have nested in the Pandanus trees for generations, and their discarded nest sticks lie in great piles around the base of the trees. In the summertime, bridled terns dance on the roof of the research station. "They're doing an Irish jig on the roof all night," Byrne says. "Tap tap tap tap tap, stop. Tap tap tap

tap tap, stop. It's like a Riverdance. It's crazy. Sometimes the sounds come in waves. Tens of thousands of birds cackling together, then it goes quiet and you fall asleep, then it comes up again!"

Other times, Albright says, shearwaters wake the researchers. The birds tend to fall asleep on the sloped roof of the research station in the evening. Then they slowly slide down the roof until they fall off and have to scramble loudly back to the top. "You just feel like you're on the edge of the Earth," Albright says. "You can't see land. You don't see anybody. It's all open to the stars. It's just—you never want to return to reality."

And all around, beyond the stars and the birds, lies the reason everyone's there in the first place: the coral.

Reef-building corals include both a host, the coral animal itself, and a symbiont, which is a type of algae called a dinoflagellate that lives inside the coral. The two work together: the algae generate energy through photosynthesis and feed on the coral's waste nutrients, while the coral offers the algae a protected home. Corals have existed, even thrived, in the ocean with this symbiotic relationship for over 200 million years.[14] A coral reef starts with a single microscopic coral polyp, often already harboring its particular algae. The polyp then starts to clone itself, eventually creating colonies of thousands or hundreds of thousands of genetically identical polyps. Each polyp in the colony builds a skeleton out of calcium carbonate—similar to the way oysters or abalone build their shells—until, over many years, they've built the beautiful structure people are familiar with. There are multiple species of dinoflagellate symbionts, and those species may be more or less adapted to different environmental conditions, which also makes the coral more or less adapted, too.[15]

Several of the most biodiverse regions on the planet rely on the coral-algal symbiosis. Where a rocky tide pool system thrives on cold, nutrient-rich water itself, the warm water of the tropics doesn't support nearly as much life. The evolution of the coral-algal symbiosis made it possible for the corals to occupy this habitat. They have the ability, working together, to create a rich ecosystem out of poor conditions. Corals are known as "foundation species," which means that the platforms, crevices, nooks, and crannies they build provide a literal foundation—a home—for hundreds of other species.[16] In every crevice created by the complexity of the calcium carbonate skeleton, eels, fish, sea urchins, anemones, octopus, worms, snails, sharks, and more find safe harbor. Coral reefs are equivalent to rainforests on land in terms of their biodiversity and ecological and physical complexity.[17] This high biodiversity contributes to the image we have in our head of what a healthy reef looks like—brightly colored, teeming with fish, structurally complex with many different shapes and sizes of coral, and perhaps an eel or octopus peering out from a dark corner.

It would be hard for a snorkeler or diver whose experience has been in popular tourist areas in the Caribbean to imagine what a healthy reef in the Coral Triangle or Great Barrier Reef looks like. The diversity of species and the interactions between them—these things fall apart as a coral reef does. Almost any place that has a lot of people nearby stresses its reefs to some degree, through pollution, fishing, or simply overvisiting. Again, the reason researchers noticed so many disease outbreaks in the Caribbean through the 1980s and 1990s wasn't because it was the only place with disease-causing pathogens or the only place scientists were looking. It was because that's where all the other stresses had weakened coral immune systems and made them less resistant when something like the *Aspergillus* arrived.[18]

Albright grew up far away from coral reefs; a competitive swimmer from an early age in Ohio, she says she was always more comfortable in the water than on land. As an undergraduate at Duke University, she had an opportunity to study abroad in Australia, and she got scuba certified in a limestone quarry back home so she could dive the Great Barrier Reef. She always wanted to be a marine biologist, she says, and gravitated early in her academic life to the beauty of coral reefs. The more she learned, however, the more she learned to see beyond the superficial beauty of colorful corals and into what she calls a "treasure trove" of different kinds of life forms interacting with each other. Like the biologists studying tide pools as a way to see an entire universe in microcosm, Albright saw in reefs all manner of fighting and cooperation, parasitism and mutualism. As a graduate student studying coral health at the University of Miami, she spent plenty of time underwater looking at reefs. Coral had started seriously declining in the same decade she was born, she told us, and by the time she was able to look for these complex relationships, the reefs were already different. "The types of relationships like corals fighting each other off with chemical defenses, we learned about that," she says. "We never saw that. You never saw two corals close enough together in Florida to see that."

Byrne says that one way you can tell you're on a truly healthy reef is in the extraordinary size, diversity, and sheer number of predators. She told us that you see sharks swimming around the One Tree lagoon, and six-foot humphead wrasse and three-foot coral trout and groupers, and "it's not just the big guys, it's from the top all the way down to the nooks and crannies." Reefs up and down the Great Barrier Reef have been devastated by outbreaks of coral-eating crown-of-thorns sea stars, but Byrne says predators at One Tree Island keep the crown-of-thorns sea stars in check. It is the seeming model of a balanced, complex, beautiful coral reef system.

Because it is so isolated, so protected, and so fully functioning, One Tree has also become a place for people who want to understand the future. Local efforts on degraded reefs around the world show that pollution can be remedied, invasive species reined in, fish recovered. Corals grow in places known for pollution and sewage spills. Reefs can even grow back eventually from small-scale blasts of dynamite by fishers.[19] But local regulation and conservation has limits when it comes to the effects of a global stressor like climate change. "It doesn't matter what the zoning is, you can't protect the reef from hot water," Byrne says.

In the early 1980s, Drew Harvell, while still a graduate student at the University of Washington, received a fellowship through the Smithsonian's Tropical Research Institute to study fish and snails that ate coral. For her fieldwork, she would explore the coral reefs of Uva Island in the Gulf of Chiriqui, about ten miles off the northwest coast of Panama. Her first dive was with Smithsonian coral expert Peter Glynn at a coral transect he had been monitoring for years. In her 2019 book *A Sea of Glass*, Harvell recounts the boat ride out to the site over a flat green sea and worrying, as an inexperienced diver, about sharks in the murky water. But as she donned her scuba gear and slipped away from the boat, her attention was quickly diverted from sharks. "Once underwater, I forgot my fears and was surprised and confused—all the coral as far as I could see was white," she wrote. "These are very shallow reefs, and we were only in about fifteen feet of water, so we gestured to Peter and surfaced. 'Why is the coral white, is it safe for us to be in the water?' we asked. He looked pretty shaken and said he'd never seen this before and didn't know."[20] A few months later, Glynn published an article describing a 10,000-square-kilometer area of whitened or dying corals on the Panamanian Pacific Coast.[21] It emphasizes just how strange the event was that in his write-up, Glynn adds quote marks around the word that today even most sunscreen-slathered tropical tourists know: "bleaching."

Scientists had known for nearly a century that when corals were stressed, the symbiotic relationships fell apart. Scientists working in labs in the early 1900s found they could induce bleaching by starving corals or by depriving them of light.[22] By the 1970s, coral biologists knew bleaching could also be caused by warm temperatures.[23] In a journal article in 1974, scientists described corals dying around Kahe Point, Oʻahu, where a power plant had discharged superheated effluent. They noted that corals exposed to temperatures two to four degrees above normal bleached but didn't die, while those exposed to temperatures greater than that died.[24]

Bleaching might happen when the coral expels the symbiont or the symbiont leaves, or worse, it begins to parasitize the host coral, but however the decision is made, the coral loses its algal partner—and incidentally also its color because that is derived from the algal symbiont.[25] The coral doesn't necessarily die, although it becomes more likely that it will. Either way, the result of the breakup is a bleached, white coral. The greater the stress, the more extensive the bleaching on a reef.

Scientists have long debated how exactly the so-called stony corals and their symbionts evolved and came to spread over the world's shallow seas.[26] Corals show an extreme flexibility in living arrangements. They thrive in their symbiotic relationships in the light-filled tropics, but they also grow on their own in deep-sea areas devoid of light, and lacking the photosynthesizing symbionts that need the sun. Many shallow-water corals form reefs, but others live alone without symbionts, and some seemingly can switch back and forth between symbiotic and solitary living. With such variation, it's possible the reef-building corals evolved on their own and much later met up with symbionts, and it's possible some of the corals and some of the symbionts met long ago and evolved together, and it's possible they've met and parted and met and parted many times over the ages.[27] However the stony corals choose to live and whatever their origin, they've all survived near-extinction before. Two hundred million years ago at the end of the geologic period called the Triassic, a rift opened in what is now the center of the Atlantic Ocean, and volcanoes bloomed like weeds.[28] Lava poured forth, and eruptions filled the sky with carbon dioxide, sulfur, and methane.

More than three-fourths of species on Earth went extinct as we entered the new age of the dinosaurs. Reef-forming, symbiont-partnering stony corals appear in the fossil record and seemingly dominated the shallow seas by the time the volcanoes blew. Like most other life on Earth, the ensuing catastrophe was rough on them. By some accounts, 96 percent of the stony corals—seventy-four out of seventy-seven genera—went extinct in what's called the End Triassic Extinction.[29] It took them nearly 10 million years to recover. So what killed them? Did they bleach, expelling their symbionts like our modern corals, as greenhouse gasses warmed the atmosphere and the ocean acidified? Or did they die from the many other catastrophic changes unfolding around them?

Researchers who study paleoclimate using corals—reconstructing the environment that the corals lived in for thousands or even millions of years—have tried to develop a way to diagnose coral bleaching in fossil specimens. Some researchers think they have found evidence that shows bleaching many millions of years ago.[30] Yet it is exceedingly challenging to draw confident conclusions from an incomplete 200-million-year-old fossil record—particularly when the dinoflagellate you're interested in doesn't fossilize. Scientists have also found it difficult to tell from coral skeletons whether they bleached in the past or not. Corals build their skeletons from a number of elements in the seawater around them, some of which change form slightly depending on conditions like pH, making them a potential record of the past.[31] These proxies, or representatives of past geochemical change, are still being studied to accurately discern past episodes of bleached versus healthy coral.

Coral skeletons tend to stay in place long after the corals have died, so some researchers have taken samples from relict coral reefs to see what they can learn about how and when the corals died. In places in the South China Sea, radiometric dating indicates that many of the corals died at the same time in regional events—perhaps tied to El Niño warming in the late nineteenth and early to mid-twentieth century.[32] Still the direct evidence for widespread coral bleaching, before people saw it happen in 1979, remains out of reach for now.

Other researchers have tried to assess the likelihood of coral bleaching more recently by reconstructing past sea-surface temperatures. In 2005, researchers used water temperature records from the logbooks of nineteenth-century and early twentieth-century ships to try to picture the oceans before widespread remote monitoring began. These reconstructions found that in the Caribbean, the Northwestern Hawaiian Islands, and the Great Barrier Reef, the water probably got too hot for corals and their symbionts several times between 1940 and 1960.[33] So why hadn't any scientists noticed before 1979? It seems likely that all the other sources of stress in a coral's life got worse. Disease, sedimentation, overfishing, acidification, and physical damage were already affecting the corals, so when the heat waves of the 1970s and 1980s arrived, the corals couldn't take it anymore.

Whatever the extent of coral bleaching in the past, the frequency and intensity of events in the past forty years is remarkably different. When Harvell and Glynn surfaced over the field of skeletal Panamanian corals in 1983, scientists who had previously documented coral bleaching largely understood it as something local. Bleaching was the result of an interesting experiment in a lab, or something you might find where human pollution was particularly acute, or where a hurricane had just passed. The first difference this time was that bleaching wasn't confined to a small area. The second was that it kept happening. Following another intense Caribbean bleaching event in 1987, a team of Smithsonian researchers set out to learn what they could from the recent past.[34] In a pre-internet, pre-email age, they sent out physical questionnaires to 694 scientists around the world. One hundred and fifty-nine of them, representing forty-eight countries, wrote back. Their eyewitness accounts showed that corals around the world had been suffering for years and uncovered new episodes that moved the clock back earlier on the first recorded bleaching events. Scientists in Puerto Rico reported seeing bleached coral skeletons after a hurricane in 1969, possibly because all the rain had made the water too fresh. In 1979, a scientist on the island of Bonaire, in the Dutch Caribbean, noticed a large bleached area that lasted for eight months. In 1980, bleaching struck parts of the Florida Keys, Puerto Rico, and Jamaica. In 1981, it happened in Puerto Rico again. In 1983, Glynn and Harvell saw the Caribbean corals, while researchers in Indonesia noted several sporadic bleaching events in the Seribu Islands, Kaimun Java Islands, and Pulau Pari. In 1986, scientists saw bleached coral in Hawai'i, Mozambique, Okinawa, Puerto Rico, Barbados, and the Bahamas.

Some of the Australian corals damaged in a 1987 bleaching event wound up in a lab at University of California, Los Angeles (UCLA), where a graduate student named Ove Hoegh-Guldberg found them and started to study them. Hoegh-Guldberg turned the causes of coral bleaching into his dissertation, establishing a direct link between temperature and bleaching. In experiments on corals taken from Lizard Island in the Great Barrier Reef, Hoegh-Guldberg found that the corals did not respond to increases in light or changes in salinity, but that bleaching followed quickly once the temperature rose above about 85°F.[35] Tropical seas don't tend to vary in temperature as much as their higher latitude counterparts, but they're also maintaining a range that's much closer to the maximum corals can tolerate.[36] Most corals live in the warmest parts of the ocean, with average temperatures in the low eighties.[37] Even as he was studying his Lizard Island samples in the late 1980s, Hoegh-Guldberg was seeing the reports about human greenhouse gas emissions warming the world's oceans. The implication was alarming. "That changed things for me," Hoegh-Guldberg tells Irus Braverman in the 2019 book *Coral Whisperers*. "I was always a conservationist. And I loved diving. But it suddenly became clear to me that the thing I loved was now threatened."[38]

Between the 1980s and early 1990s, coral researchers around the world had started to piece together the links between coral, ocean temperature, and bleaching. When Harvell and Kim saw corals turning white in the Florida Keys in 1998, they both knew what was happening. But somewhat like the first time Harvell had seen bleaching in Panama, it was still hard to understand the full scope of what they were seeing. Because until that year, scientists had never seen every ocean on Earth turn hot at once.

In April 1997, sensors picked up a massive slug of warm water spreading across the equatorial Pacific, the hallmark of an El Niño.[39] In places, it was 3°C to 5°C above average. Over the next year, the warmth sloshed up along the North American coast and down along South America in the east, and into the Indian Ocean in the west. It pooled in the Coral Sea off Australia. The force of an El Niño causes winds to weaken over the Caribbean, and temperatures soon started rising there as well.[40] Just about everywhere on Earth where coral reefs live felt the heat.

What Harvell and Kim saw in the Florida Keys would become known to coral researchers as the first "global bleaching event," defined as simultaneous bleaching covering 100 square kilometers in all three ocean basins—Indian, Pacific, and Atlantic, including the Caribbean and the Gulf of Mexico.[41] As far as anyone knows from modern records, one had never occurred before. Corals that had lived for centuries through every kind of ocean condition to date turned white and gave up in the face of this relentless new heat.

"For coral people, it was horrific," Harvell told us in 2018. "Because of the bleaching, [and] all these infectious disease outbreaks that were related. That kind of started as a recognition that the sustainability of coral reefs was desperately in danger. That's the point at which we said, 'Oh my god this is going to be terrible. It's going to get worse.'"

The Caribbean had become an important testing ground for understanding tropical disease outbreaks and understanding how corals and other organisms were developing resistance and even chemical cocktails with antibacterial properties to confront these outbreaks.[42] "The disease outbreaks in sea fans, elkhorn corals, and staghorn corals that occurred in the 1980s and 1990s, along with rapidly declining live coral cover, lit the fire under scientists and policymakers," Harvell writes in *Ocean Outbreak*. She goes on to describe the "disastrous effects of the 1997–1998 El Niño warming event on coral health and the dawning realization that many national and local economies were dependent on healthy reefs."

Harvell, Kim, Smith, and several other disease specialists sounded one of the first alarms about the warmer world and the spread of marine disease in a paper published in 1999.[43] Although it wasn't clear yet whether it was more disease or just more people noticing it, Harvell and Kim wrote that the entire community structure of the Caribbean had changed due to mass die-offs in "plants, invertebrates, and vertebrates." Outbreaks of bacterial and fungal diseases in the Indo-Pacific had killed coralline algae at an unprecedented scale. Marine mammals were getting sick and dying in the North Atlantic. Seagrass, oysters, and sea urchins had suffered epidemics in temperate waters worldwide. The 1997–1998 El Niño exacerbated viral outbreaks in oysters in the Gulf of Mexico and in Mediterranean monk seals in Mauritania. "Contributing to the emergence of new diseases," the paper concludes, "would be a long-term warming trend, coupled with extreme ENSO [El Niño-Southern Oscillation] events and human activities that have modified marine communities."

Also in 1999, Hoegh-Guldberg published a paper proposing that, based on temperature alone, corals would disappear by 2050. "Most information suggests that the capacity for acclimation by corals has already been exceeded, and that adaptation will be too slow to avert a decline in the quality of the world's reefs," he wrote. "The rapidity of the changes that are predicted indicates a major problem for tropical marine ecosystems and suggests that unrestrained warming cannot occur without the loss and degradation of coral reefs on a global scale."[44]

Once the connection was made, the projections suddenly became clear. The world was warming, disease and other stressors were spreading, and the corals were rooted in place. The Caribbean had suffered badly in 1998, but it had also suffered occasional bleaching and disease outbreaks for decades. This time, even remote, protected corals on the Great Barrier Reef had been damaged. Most reefs bounced back quickly, but it was a warning sign. The question started to become, How quickly would the oceans continue to warm? In 2010, another widespread bleaching event struck, this time as part of a weak El Niño, with many more scientists watching. Then, in 2014, the third global bleaching event started.[45] This one did not end as quickly. For three years, heat waves swept across the land and through the ocean. Nearly 75 percent of coral reefs worldwide were exposed to temperatures that could cause bleaching. Nearly 30 percent were exposed to temperatures that could cause them to die.[46] More than 80 percent of corals in the northern sector of the Great Barrier Reef were severely bleached, and almost none had escaped. Australian coral scientist Terry Hughes, who had monitored study areas on the Great Barrier Reef since the 1990s, posted online that he showed the results of the 2016 aerial survey to his students, "and then we wept."[47]

Although the third global bleaching event officially ended in 2017, bleaching has continued. In February 2020, scientists recorded the highest sea-surface temperatures in Australia since they started keeping records in 1900. One quarter of surveyed reefs were severely affected by bleaching, and 35 percent had some bleaching. Warm water crept southward to affect even sections of the reef like One Tree Island that had escaped the worst of the heat in 2014–2017.[48] In 2022, for the fourth time in seven years, another mass bleaching swept through.[49]

In 1975, the Australian government created the Great Barrier Reef Marine Park to protect many areas of the reef from most forms of fishing, mining, and geological exploitation.[50] Marine protected areas were designed to limit specific uses of a particularly ecologically or culturally important seascape. Over time, as climate change emerged as a global threat, the theory behind that protection has evolved into an idea that removing human-caused stress on a natural ecosystem would increase its resilience to processes like global climate change and bleaching. Until recently, that might have seemed plausible. If bleaching events were a decade or more apart, well-protected reefs could potentially recover. Many Australian reefs recovered from the 1997–1998 El Niño and global bleaching event. Now, as these events become more and more frequent, it is not clear that even the most protected corals will recover.[51] The third global bleaching event, coming just four years after the second and lasting for an astonishing three years, marked for many coral scientists the switch to a different world.

———

Brown University climate scientist Kim Cobb studies corals in one of the most remote places in the world, Palmyra Atoll. Go to Hawai'i, then go south about 1,000 miles into the middle of the Pacific Ocean, she told us. Some of the most pristine coral reefs in the world live there. Twice a year, Cobb and her graduate students go to study the reefs. She arrived in November 2016, at the tail end of the recent global bleaching event, as one of the strongest El Niños ever recorded had superheated the tropical Pacific. Cobb says she was emotionally prepared for half the corals to be dead.

They arrived on the island, and everything was going perfectly. No equipment mixups, no visa issues, perfect weather, calm seas. They took the boat to the farthest-flung reef. "I remember getting in the water, and looking down, and thinking, 'It's too deep here. We're way too deep. I can't see anything,'" she said. She wondered if maybe they had gone to the wrong place. "Halfway down I started to recognize the lay of the reef topographically. This is the reef. There's just nothing left. As you got closer it was just straight brown red algae covering the entire reef. A centimeter or two kind of wafting in the current. I was . . . in no man's land. It was like an out of body experience. Some place you've called home for so long, spent hours diving that site for so long, it was like diving on the moon, that was how foreign it was."

They got back in the boat, stunned, and went to a site closer to the research station. It is a site "like our backyard," Cobb says, where she has been diving probably 100 times. She had worked on those corals for twenty years, since she was a graduate student. She knew them individually. They, too, were gone. "They were just wiped," she said. "Dead. That's when it hit me, like a close family member at that time. It's like a betrayal. 'You too? Are you serious?' I felt angry. I felt angry at that site. Shocked at the first site and angry at the second. Angry at the system. After the anger, I cried on that dive. I've never cried on a dive before. I shed tears into my mask. That's when the full depth of the realization sunk in. I knew as a scientist that if that site succumbed, the entire reef, there would be no place that escaped. I knew everything would be gone. There would be no silver lining here."

A month after she had returned to the United States, Cobb gave a speech at a climate protest outside the American Geophysical Union conference, an annual meeting in San Francisco of tens of thousands of the world's top earth scientists.[52] Standing along with other people in white lab coats, next to a protester holding a sign reading, "Ice Has No Agenda, It Just Melts," Cobb addressed the scientists in the nearby conference to ask, as people who care about the world, what they were waiting for. She talked about how it had been hard for climate scientists to watch the planet setting temperature records and mentioned her dying corals. "We have for too long as scientists rested on the assumption that by providing indisputable facts and great data that we are providing enough to counter the forces against science," she said. "And obviously that strategy has failed. Miserably."

"It was completely ad-libbed," she told us several years later. "Just spoken from the heart, realizing how profoundly I felt I had failed, as a scientist perhaps. How profoundly as an extension the science institution has failed." A few years later, Cobb testified before the U.S. House of Representatives Committee on Natural Resources that "2016 was my wakeup call."[53] "Can we unlock something from this train wreck?" Cobb asked us. "My practice of science may never be the same, and my approach to my life may never be the same."

It hurts to witness the transformation of a beloved habitat, to watch a coral reef die. It would be easier to turn away. For coral scientists, it is perhaps the most life-altering part of the last twenty years that the threats to coral reefs have not abated with more knowledge. Study after study shows that, to save the reefs, carbon emissions must come down, and through every bleaching event and every publication, emissions keep going up. Compare the major dates in coral bleaching science to the atmospheric carbon dioxide record observed by National Oceanographic and Atmospheric Administration (NOAA) on Mauna Loa. First coral bleaching observed in 1983; carbon dioxide roughly 340 parts per million (ppm). First widespread global bleaching in 1997–1998; carbon dioxide 360 ppm. Second global bleaching event in 2010–2011; carbon dioxide above 380 ppm. Third global bleaching event in 2014–2017; carbon dioxide near 400 ppm. Fourth global bleaching event in 2020; carbon dioxide 410 ppm. In 2022, as another bleaching event spread across Australia, carbon dioxide concentrations hit a record high: over 420 ppm.[54] Thousand-year-old coral reefs are dying everywhere we look, and human emissions continue ever-upward as people, governments, and economies refuse the hard choices to reduce them. The dangers, instead of diminishing, have compounded. And warming and disease form only part of the new threat to corals.

In 1998, just as the first global bleaching event was starting to recede, a postdoctoral scholar at the National Center for Atmospheric Research named Joanie Kleypas found herself sitting at a meeting on ocean chemistry. A colleague asked Kleypas, an expert in how coral reefs are affected by global change, if anyone had predicted how reef-building corals would respond to changes in chemistry associated with rising carbon dioxide concentrations in the atmosphere. The answer was: not really. Yet at the coarsest level, it's not a terribly difficult calculation. The ocean absorbs roughly one-quarter to one-third of human carbon dioxide emissions. That carbon dioxide, as it dissolves into the water, makes the pH slightly more acidic, increasing a particular type of carbon called carbonic acid and in turn decreasing the relative amount of a different type of carbon called carbonate.[55] While the relationship between the coral and its symbiont has fascinated people for millennia, so too, has the ability of corals to build a hard, protective skeleton out of calcium carbonate, and particularly a mineral form of it called aragonite. Change the amount of carbonate available, and you change how readily the corals can build their skeletons. When Kleypas put pen to paper at that meeting, scientists already knew that lower carbonate concentrations limited coral growth.[56] They had described corals living in more acidic seawater in the present and studied ancient corals from more acidic times in the past. They also had in hand the projections of the second Intergovernmental Panel on Climate Change (IPCC) showing rising human carbon dioxide emissions. Kleypas's calculation meant, essentially, putting it all together: if we have this amount of emissions, this amount of carbon dioxide in the ocean, and this amount of carbonate ion removed, what does that mean for corals?

Kleypas was a lifelong ocean person. She told us that her first memory was her mother teaching her to swim when she was two. She had always known she wanted to work in the ocean. She grew up in a refinery town in Texas and vividly remembers the contrast of the polluted landscape with her childhood as an animal lover, "always trying to save something." She remembered returning from the beach once, as a little girl, and a fish falling out of her swimsuit. She ran tap water over the flopping fish, trying unsuccessfully to keep it alive. "I still feel bad about it," she said.

Science offered Kleypas the opportunity to keep trying to understand things. "I didn't go into this work because of climate change," she told us. "That sort of happened after I was in college. It just sort of falls on your shoulders." After graduating with an undergraduate degree in marine science from Lamar University in Beaumont, Texas, she won a Fulbright Scholarship to study at James Cook University in Australia, where she first started working on coral reefs. Coral combined her interest in marine biology with her interests in chemistry and geology. When she got her PhD, she returned to the U.S. Geological Survey and National Center for Atmospheric Research to model why reefs were where they were.

While the meeting went on around her, Kleypas had the perfect background to rough out the back-of-the-envelope calculations of the relationship between the health of coral reefs and the chemistry of the ocean. Still, the result stunned her. The reefs were in real trouble. "I felt physically ill," she told us. A year later, Kleypas published the first paper that documented the impacts of changing ocean chemistry on coral reefs, a study titled "Geochemical Consequences of Increased Atmospheric Carbon Dioxide on Coral Reefs."[57] The knowledge was beginning to trickle out—Kleypas was one of several scientists thinking about the same problem around the same time. Jim Barry, whose paper on intertidal species helped reshape the way marine biologists thought about warming, told us that Kleypas's 1999 paper was a similarly dramatic turning point for the marine ecology community. Ocean ecosystems, places that for centuries even many scientists had long considered too big for human influence, were being affected by human activities. The burning of fossil fuels was creating a footprint that no ecosystem, no matter how far from land, no matter how protected, could fully avoid. The chemical changes to the ocean that occur through absorption of carbon dioxide affect the tropics and the poles. It happens in shallow water, and it trickles into the deep sea: as universal and inevitable as the laws of chemistry themselves. "A coral reef represents the net accumulation of calcium carbonate ($CaCO_3$) produced by corals and other calcifying organisms," Kleypas and colleagues wrote in the first line of the paper. "If calcification declines, then reef-building capacity also declines."

Kleypas told us that no one really wanted to be the first author on the paper because the claim was so significant and the stakes so high. She and her coauthors felt a great need to be cautious, to avoid sounding an unnecessary alarm. And yet the result was the result. "If you see a meteor heading at the earth," Kleypas said, "you have to do something."[58]

Kleypas told us that she hoped repeatedly over the years to be proved wrong. At first, she wasn't sure about all the chemistry, so she would consult other chemists, each with a different expertise. Every time, she said, she hoped they would find a mistake. Every time, the consulting chemists would return a few days later and say, "I agree." After the paper came out, scientists would write to her and say they were sure she was wrong and planned to prove it. Despite the embarrassment it would mean, Kleypas said that, deep down, she still hoped they would. None ever did.

Five years after Kleypas's paper in *Science*, Ken Caldeira and Michael Wickett coined the term "ocean acidification."[59] Scientists would learn much more about it over the coming decades: about the many complexities in the chemistry of the ocean and the way it interacts with biology, and how ocean acidification overlaps with other stressors for reefs and other systems. There are now more than 250 scientific papers published every year on the topic of ocean acidification. Federal agencies have programs that specifically fund ocean acidification research. But as many researchers we interviewed noted, simply adding knowledge hasn't reduced the threat. In speaking with us, Kleypas highlighted what she sees as a need for broader changes to the scientific endeavor itself: greater integration with policymaking, prioritized funding, and more collaboration. "This is bigger than our individual careers," she told us. She finds a glimmer of hope for reefs by taking breaks from her continued research on coral adaptation and management under climate change to do direct restoration work.

Kleypas concluded her 1999 paper with a list of uncertainties, many of which have now been investigated. How exactly does a coral's ability to calcify relate to the carbonate saturation of the surrounding water? How much would the effects of acidification differ from species to species? Would some locations be different than others? Could some species adapt to gradual changes in chemistry? She and her colleagues concluded that, given the math they had done and observations already on hand, it seemed very likely that ocean acidification wasn't just a problem for the future. The "aragonite saturation state," a prediction of whether the environment allows animal shells to grow or dissolve, had declined over the past century. "Net calcification," the paper concludes, "has probably already decreased on some reefs." By the time Kleypas and her colleagues had defined it, the changing ocean had already arrived.

Over time, Maria Byrne says, the researchers who have come to One Tree Island have shifted dramatically from exploratory science about the nature of coral reefs to investigations of the effects of rising temperatures and changing chemistry. While tide pool researchers watch many species move through their study area, coral researchers come to One Tree Island to see what happens when a remote coral reef, strictly protected for fifty years from any human use except research, bathes in an ocean-scale change. In 2014, a group of scientists who had been working on One Tree Island for over a decade wondered what would happen if they could temporarily turn back time on the reef and understand how the coral used to function. Rebecca Albright and Ken Caldeira led this team's field excursion, which included assistance from seventeen other people, in the lab and the field, to construct a temporary time machine.[60]

Albright and her collaborators planned to take advantage of One Tree Island's three distinct lagoons, each surrounded by a coral wall that is fully exposed at low tide and submerged at high tide. The scientists picked a section of the wall separating Lagoon One and Lagoon Three, where anything they added would flow in a predictable pattern over the reef and into Lagoon Three. In September 2014, they installed a 4,000-gallon tank on the edge of the coral. Then, for fifteen of the next twenty-two days, they mixed sodium hydroxide with color-dyed seawater in the tank and pumped it over the coral reef for an hour. The solution changed the water chemistry to become more basic, temporarily restoring the conditions to those in the oceans before the beginning of the industrial age in 1850. At monitoring stations along transects across the reef, they measured how much carbonate the corals took up and how quickly they grew. The reef flourished, growing—calcifying hard skeletal parts—about 7 percent faster than the days when no manipulated water was added. In a paper published in 2016, and in an echo of the speculation at the end of Kleypas's paper two decades before, Albright and her collaborators wrote that they had found "ocean acidification may already be impairing coral reef growth."

With some of the same team of researchers, Albright returned to One Tree Island two years later to run the reverse experiment. What if, rather than turning back time and watching the reef grow, we sped up time and watched what might happen in the future? This time, the team prepared a huge tank of water that had higher than normal carbon dioxide concentrations in it—water chemistry representing the reef conditions in 100 years. As they watched the dyed, acidified water flow across the reef, they measured the decline in reef growth—a 40 percent decline relative to today.[61]

Both of Albright's studies took science that had previously only been done in the lab out into the field. There are very few places on Earth where scientists could design such a perfect field-lab experiment as One Tree Island. But the remote, pristine nature of the site also meant something much bigger about the ability of protected areas and remote locations to help corals escape the future. In Australia, on the Great Barrier Reef, in the world's most iconic marine park, the corals still declined. As Byrne later told us, there's no carving out a local exception for an ocean of hot, acidifying water. Researchers who once saw the Caribbean as degraded now look to it as a model for what will happen everywhere. With no sign that the train is slowing, Albright has turned to looking for ways to connect coral spawning in the lab to restoration efforts in the field, in the hope that they can help the corals brace for the impact, and to working with community organizations to smooth restoration efforts around the world.

The tension between local community conservation action and inescapable global change is constant in ocean science now. It shapes the work of every person in the locations described in every remaining chapter in this book: kelp forests, shellfish beds, fisheries, the open ocean, the poles, the deep sea. But coral scientists got there early. The psychological effects of such sustained pressure on the community of coral researchers have been profound.

When we talked to her in 2018, Drew Harvell was studying a viral outbreak in sea stars from Friday Harbor Laboratory in Washington. We asked her, What lessons could she take from having worked around the world in so many kinds of ecosystems? Harvell answered, "There's trouble everywhere." But the distance also gave her some perspective on the particular sorrow of coral researchers. "I feel lucky that I'm not strictly a coral biologist," she told us. "Because there's no relief for those folks. At the last coral meetings I was going to, people were crying in the hallways."

In the 2018 book *Coral Whisperers*, ethnographer Irus Braverman interviews hundreds of coral scientists around the world and describes them oscillating between hope and despair and between traditional and interventionist remedies.[62] She argues that they are sharing an experience with many other people who care about the natural world. The faster the climate changes, the more pressure everyone will feel. "*Coral Whisperers*," Braverman writes, "is a book about bearing witness. Specifically, it is a testimony of how coral scientists are coping with the unfolding coral crisis of our time: the most catastrophic period for reef-building corals in human history, and for the preceding sixty million years."

Braverman argues that the pressure of watching their subjects die, continually calculating the grim odds for the future, and having to narrate that death to an often indifferent public and often hostile politicians has traumatized the coral research community. Being on the front lines of the decline of one of the world's most important ecosystems, and arguably one of the most popular, generates a unique toll. Yet coral scientists have chosen to continue to observe, record, and experiment as calamity takes hold in systems they cherish. As Braverman describes, they haven't just watched; they have narrated the plot of this drama for the world to see, and they are at the forefront of suggesting solutions for the future of corals. The cost of such work in terms of mental health has itself been the subject of the feature film *Chasing Coral* and national magazine articles.[63] Braverman's book is framed around coral scientists as an example system of people trying "to understand their world both individually and as a community."

The same fears and frustrations face people who love monarch butterflies or tropical frogs or bats or giant sequoias. Some coral researchers talk about lessons they have learned from the world of terrestrial silviculture, which has responded to the movement of trees and death of forests on a warming planet by actively replanting forests where they'll thrive in future climates. At one point in *Coral Whisperers*, Braverman writes that coral has replaced the polar bear as the "poster child" of climate urgency. The heart of the debate about the future, she argues, is whether people even *can* help restore corals or whether any effort spent on corals that's not spent cutting emissions is simply a waste of time. Planting coral gardens might sound like an obvious solution, but the terrestrial analog shows some of the challenges as farmers around the world shift what they plant in the face of the changing climate.[64] On the question of human intervention for coral reef resilience, the community of coral scientists has split deeply and acrimoniously.

One group, many of them longtime observers of the Great Barrier Reef, argue that restoration is futile. The Great Barrier Reef has been given the best protections and the best chances that conservation policy can grant it, and because we haven't cut emissions, it's dying just like every other coral reef on Earth. There is no point spending limited resources restoring coral reefs, they argue, when they are just going to die in the next heat wave. Braverman writes that the group finds restoration efforts "Pollyannaish" and "unscientific"—a way of avoiding the reality of the drastic changes that are needed.

The other group argues in favor of a variety of interventions that might help coral: from local protection to replanting efforts, to genetic modification. Braverman ends *Coral Whisperers* with a transcript of a question-and-answer session with Ruth Gates, who led coral restoration efforts at the University of Hawai'i and died of complications from an operation in 2018.[65] Gates pointed out that nearly all the coral died in Kāne'ohe Bay in Hawai'i after decades of raw sewage flowing into the bay and that, once the sewage was diverted, the coral returned and flourished. "I am realistic about where we are and about what the trajectory of change for coral reefs is," Gates told Braverman. "For the most part, they are almost all declining. That means it's not sufficient to step back and wait."

Albright told us that she thinks it shouldn't have to be an either-or debate between cutting emissions and restoring coral. Funding is limited, she says, but it's not *that* limited. "It's obvious the only solution is cutting fossil fuel use," she says. "That's been the only solution for many decades but we haven't done it. The reason we haven't gotten political will and social change isn't because one million dollars are going into restoration. I don't see them as mutually exclusive."

Most corals can clone themselves, or, if the conditions are right, a piece of coral can break off and regrow on the seafloor, much like a garden cutting. Many attempts at reef restoration focus on planting coral cuttings in appropriate places and watching them grow. But the majority of corals also reproduce sexually, broadcasting clouds of gametes into the water that mix with each other and make new corals that drop to the seafloor. As with all creatures, this mixing of genes means greater genetic diversity, and in a time of major environmental change, that diversity can mean the difference between survival and extinction. Unfortunately for scientists who would like to encourage more spawning in the lab, corals are like pandas: notoriously finicky about the specific conditions that cause them to reproduce. On the Great Barrier Reef, more than 100 species spawn simultaneously one time per year, for maybe a few days, in a mass event determined by water temperature and quality of moonlight. If it is cloudy or if there is light pollution, the corals delay or call it off. One study found that coral gene expression changes based on the phase of the moon. Corals, researchers note, do not have eyes to gaze up at the starry night sky. Their timing seems to be based on photosensitive molecules that help them synchronize with the light around them.[66]

Scientists have been working to figure out how to predict and control the process of coral reproduction. With animals so deeply sensitive to circadian rhythms, that task has been exceedingly difficult. Researchers must often wait for the spawn to happen, then collect eggs and sperm they rush back to mix in a nearby lab. Over the next few days, embryos form and turn into free-swimming larvae that then need a place to settle and grow. Only about 10 percent survive. "Improving lab-raised polyps' chances of survival is the biggest remaining technical challenge for assisted recruitment," writer Michelle Nijhuis concluded in a *bioGraphic* magazine piece on Caribbean efforts at growing corals.[67]

Albright and her colleagues at the California Academy of Sciences have built on work by Jamie Craggs at the Horniman Museum in London to push the technology a step further, raising corals in tanks that closely mimic ocean temperature and light conditions, so that the corals choose to spawn in the lab. In 2020, Albright did an Earth Day live stream as colonies of *Acropora hyacinthus*, replanted in the tank from fragments collected in Palau, spawned behind her. The corals have been tricked for the convenience of the researchers: the lighting in the entire room is carefully controlled to run day and night cycles reversed from reality so that the researchers can work during normal office hours. In a dark room in the middle of the day, lit by the red glow of the custom-programmed tank lights and wearing a face mask dotted in ladybug prints, Albright gathers the coral gametes in nets to mix and raise in a tank next door. "Once a year, that happens," she says. "So you've just witnessed something pretty special."[68]

Acroporid corals are widespread in the world's oceans and an important foundation on most coral reefs. The California Academy of Sciences uses its lab-raised corals for research and restoration science. Scientists still have fundamental questions about coral basics so the Academy collaborates with scientists nearby at Stanford University, the University of California, and the NASA Ames Research Center. The *Acropora* could become a model organism for corals, a species or two that's so studied and so well understood that it becomes a proxy for many others. Albright's group wants to advance the understanding of raising corals in a lab through to the breeding stage and have the same corals fall into the same annual spawning rhythms they would in the wild, with the hope that it becomes easier in the future for more land-based places to grow corals for restoration and planting. In the video of her live stream, Albright says that the corals spawning behind her had been collected in Palau months earlier but that other corals in the tank had been there for more than a year, had spawned the previous year, and seemed to be on the road to trying again. (Later, she told us, the second year had also been successful.) Albright told us that Academy scientists are studying the genetics of each generation of corals, from the parents onward, to try to learn more about what makes corals thrive in different situations. In 2021, her group generated a much higher resolution record of the *Acropora hyacinthus* genome based on the spawning corals in the lab, which Albright says allows them to answer even more rigorous scientific questions.

One hope is that eventually corals raised in labs on land and bred for heat tolerance and disease resistance might be planted in degraded areas to secure a future for reefs. In *Coral Whisperers*, Ruth Gates spoke about using lab experiments to increase coral performance—to speed up evolution, in some sense, by identifying and selecting traits that will benefit corals as ocean conditions continue to warm and acidify, similar to the crossbreeding that has occurred in agricultural crops.

So far, the research on improving coral resilience has been mostly local. Albright and the California Academy of Sciences partner with an international nonprofit group called SECORE (SExual COral REproduction) International that works to restore genetic diversity to damaged Caribbean reefs by facilitating coral sexual reproduction. SECORE has also developed better methods for planting corals and has researched ways to make it possible to sow young corals from the surface, like scattering seeds. In *Coral Whisperers*, scientists disclose some recent successes: in Bermuda, regulations that reduced fishing and pollution resulted in increased coral cover following restoration; restoration and coral planting efforts in Belize have increased coral cover on one two-acre reef from less than 6 percent to over 50 percent over the past decade.[69] Marine biologists have identified corals that survived repeated bleaching events on their own in some of the worst-hit areas of the Pacific and Great Barrier Reef.[70] In 2016, in a small area near Heron Island in the Great Barrier Reef, scientists planted dozens of colonies of lab-raised corals. In December 2021, five years after they were planted, the corals spawned for the first time.[71]

Automation also helps scale coral restoration quickly. Researchers can now use three-dimensional printers to make fake seafloor for corals to attach to. Others have explored using robots and artificial intelligence to mass-manufacture coral skeletons, potentially saving the polyps the trouble of having to secrete all that calcium carbonate and build up slowly. Robots can also be trained to do the delicate restoration work of planting corals underwater, removing the time limits faced by human scuba divers.

Hanging over all the restoration, however, is the recognition that the scale of success has been small so far compared to the problem, and no matter how cared for they are, there are some environments where a coral simply can't survive. All the effort to raise corals in the lab and sequence their genes and design new ways of restoring reefs is what coral scientists often refer to as a form of triage, appropriate for an emergency room setting. It buys time. It doesn't buy clean air or water.

When it comes to Caribbean coral declines, scientists still debate the relative importance of local factors and global climate change. Ove Hoegh-Guldberg argues that the Caribbean offers a preview for the future of the Pacific. "The Caribbean definitely felt the impacts of climate change first. It was hidden among ideas that this was about local pollution, but if you look back you can see that the disease outbreaks were probably triggered by warmer-than-normal temperatures," Hoegh-Guldberg says in *Coral Whisperers*. "But the key point is that the Pacific is now showing impacts on that scale. We're probably about twenty years behind the Caribbean." Other scientists agree that warming damaged the reefs badly but suggest that rising pollution, habitat destruction, and overfishing harmed Caribbean corals before temperatures warmed enough to cause significant problems.[72]

In the era before coral bleaching went mainstream, coral scientists might have dismissed the Caribbean as uninteresting because of its obvious damage. Now the same qualities make it appealing to study. The story of the Caribbean—the disease, human impacts, bleaching events, warming trends, and restoration efforts—provide a vivid picture and an early warning of what reefs around the world may face. As scientists work to save coral reefs for future generations, success stories in the Caribbean may prove to be important for understanding the recovery and resilience of reefs elsewhere.

Nearly forty years ago, a rampant and still somewhat mysterious disease outbreak swept through populations of a Caribbean urchin called *Diadema antillarum*. The urchin gets its common name, the long-spined urchin, from its exceptionally long and beautiful spines—imagine a bundle of knitting needles poking out through a crevice in a rock. *Diadema* matters quite a bit to a healthy coral reef. It is a voracious consumer of algae, and algae competes with space for corals. In the battle between coral polyps that want to settle the seafloor and algae, both seeking more space on the reef, the urchins keep the algal turf in check and prevent it from smothering the coral. When urchins live in high densities in a healthy reef, they scrape away the growing algae and also remove some of the calcium carbonate skeleton of the coral itself, giving them another label: "bioeroder."

The first urchins started dying at the Atlantic mouth of the Panama Canal in January 1983. The die-off moved quickly and soon spread from the Bahamas in the north to Venezuela in the south.[73] In a little more than a year, it affected all known populations of *Diadema* in the Caribbean, killing between 90 and 100 percent of urchins in every place it touched. The cause remains unknown, although from the way it seemingly followed oceanic currents, researchers suspected a waterborne pathogen. Even in a place that has seen a lot of marine death, the *Diadema* die-off "stands out for its sudden onset, intensity, geographic extent, and impact of the loss of a single species on entire ecological communities," a Smithsonian Tropical Research Institute marine scientist wrote. The delicate balance between algae and coral broke down, and reefs became carpeted with algae, diminishing the diversity of fish and invertebrates on the reef and affecting tourism and fisheries in the region. This process, accompanied by the loss of parrotfish, another key consumer of algae, because of overfishing in the Caribbean, fundamentally changed the function and resilience of these reefs.

The next forty years have been a somewhat unsatisfying story, in a way that may reveal a lot about where the world is headed. The *Diadema* never fully recovered. Recent estimates suggest they are only at about 12 percent of their one-time numbers, although that's a modest increase from the days immediately following the die-off. The reefs they once patrolled also never recovered. As researchers are finding out around the globe now and as the effects of climate change become more apparent, it is often very easy to break things and very hard to put them back together again.

In specific places around the Caribbean, however, the urchins bounced back, and their presence seemed to exert a rebalancing presence on reefs degraded for a variety of reasons. In Discovery Bay, Jamaica, damage from overfishing, pollution, and two hurricanes reduced coral cover from 90 percent to 5 percent between 1950 and 1990. Algae took over in its place, carpeting everything from the shore to a depth of 100 feet. For some still unknown reason, however, the urchins in Discovery Bay made a comeback in the mid-1990s. A small population in 1992 turned into a normal population in 1995 and persisted as an abundant population through the early 2000s. The coral bounced back on the same time line. By 2001, juvenile coral covered as much of Discovery Bay as it had in 1982, before the urchins disappeared.[74] Taking the next step, Stacey Williams, an ecologist at the Institute for Socio-Ecological Research (ISER) Caribe in Puerto Rico, grew urchins in a lab at the University of Puerto Rico, Mayagüez, and released them into four sections of reef in 2018. In just two months, the urchins rapidly reduced the algal cover in her study sites. Although the experiment didn't last long enough to measure a change in coral cover, Williams anticipated one. She saw three juvenile corals in her study site, once smothered in algae, now growing clean. "The likelihood of these coral recruits surviving has increased as they do not have to compete for space with fleshy macroalgae," Williams wrote in 2021.[75]

The *Diadema* die-off reveals what sudden, rapid global change can do. It also has begun to show what local, small-scale efforts can achieve in the face of that existential threat. And it shows the challenge of trying to understand the ocean. In 2022, a new *Diadema* die-off had begun, first observed in the U.S. Virgin Islands and documented via volunteer community science efforts on several islands of the Caribbean.[76] The extent and impact of this most recent die-off is still being studied.

What do you call this in-between zone, when a living organism is traumatized by a process beyond any individual person's control, yet it persists and, in places, fights hard to maintain in the face of adversity? It's a question the rest of the book will explore.

CHAPTER 3

THE FOREST

In a hundred years none of these trees will be here. No object thick with pitch to make the mind reflect. And if we do not call them by their names we will lose not only the trees themselves but also all trace of their having ever been. Looking at the bare tupelos at the farthest edge of Jacob's Point, I am reminded of something John Bear Mitchell said when my students asked him how the Penobscot people of Maine have responded to centuries of environmental change. "Our ceremonies and language still include the caribou, even though they don't live here anymore. . . . The change is in how we acknowledge them." His response surprised my students. He seemed to be saying: learn the names now, and you will at least be able to preserve what is being threatened in our collective memory, if not in the physical world.

—Elizabeth Rush, Rising

It's easy to imagine the bright colors of a coral reef fading to bleached white, the mosaic of reef fish disappearing, the sun dappling a barren seafloor. Yet warm water sloshing around the world's oceans has reshaped all kinds of habitats. As Australians mourned the bleaching of the Great Barrier Reef, scientists warned of a contemporaneous die-off of 90 percent of Australia's southern kelp forests—a disappearance, they said, with economic and ecological costs proportional to the bleaching of the coral reefs. In a nod to the importance of the kelp and as a way to try to draw attention to its vulnerability, Australian conservation scientists have recently begun to call their kelp-dominated southern coastline the Great Southern Reef.[1]

Kelp forests wrap around the shorelines of the world, hugging about one-quarter of the world's coastlines. Move beyond the tide pools and reefs into the "subtidal" on any continent except Antarctica and you'll find the brown seaweeds that are almost universally considered the world's underwater trees. "Literally millions of tiny animal residents may occupy a few strands of kelp, creating an underwater metropolis that rivals a tropical jungle in color, diversity, and sheer abundance of life," explorer Sylvia Earle wrote of the kelp forests off the Los Angeles coast in 1980.[2] Dive into them and, as scuba divers worldwide have found, the forest parallel bewitches the senses. Streaming vines stretch toward the sunlit surface, where the leaves, called blades, spread in a vast canopy. Fish dart through patches of shade and meadows of light on the ocean floor, between the rocks where the kelp anchors itself with a rootlike holdfast. The currents that rush fast over sandy seafloor calm in hushed pockets in the depths of the forest, where animals find shelter from the turbulence outside.[3] In the summer, as the water warms, some kelp species change color and drop their blades. "By late August, yellow and orange blades drift slowly down, reminding me of a New England autumn," a marine scientist wrote in a 1972 *National Geographic* article headlined "Sequoias of the Sea."[4] The interplay of sun and shade filtering down through the pillars has also called more than one person to compare them, like their terrestrial forest counterparts, to cathedrals. "I sometimes got lost in the kelp," Tasmanian Aboriginal diver Rodney Dillon told *Washington Post* reporter Darryl Fears in 2019. "I would lose concentration from catching food and go to look, sort of sky-gaze, at the beauty of the light coming through."[5]

There are 112 species of kelp worldwide, with an enormous diversity of shapes and styles. The iconic large kelps that branch into surface canopies grow in Australia, North and South America, East Asia, and South Africa. Smaller, ecologically important but less-noticed kelps grow from the seafloor without reaching the surface. One thing that unites all kelps, however, is a sensitivity to warmth. The typical upper temperature range for Arctic and temperate kelps is around 70°F; the lethal temperature for the most warm-water tolerant of kelps is about 75°F.[6] They live in the tropics only under specific conditions where local upwelling cools the water, often in deeper subsurface waters.[7] Scientists think kelp originated 100 million years ago in the cold waters off the coast of northern Japan. Based upon kelp genetic studies, a southward migration seems to have happened three times in the last 35 million years, and once kelp has crossed the Equator from south to north.

Anchored to its rocks, kelp may seem an unlikely candidate for the poleward march of marine species that we may see in other environments as temperatures warm. Yet the terrestrial analogy holds again: on land, trees drop seeds just far enough from their bases that forests migrate over millennia following glaciations and temperature change; scientists worry now that, as the world warms more quickly than it ever has before, many forests have run into evolutionary dead ends from which they will move too slowly to escape.[8] The "sequoias of the sea" seem to be in similar trouble. Kelp "seeds" are actually free-floating spores that can move as much as several miles from their origin. But as the oceans warm, kelps are already showing declines in places where temperatures have neared their tolerance limits. In California—where giant sequoias might disappear without human intervention—bull kelp forests in the northern part of the state and giant kelp in southern California and Baja California, Mexico, declined by 90 percent in the marine heat wave between 2014 and 2016.[9] In Australia, extreme heat waves led to multimillion-acre wildfires in 2019 and 2020. "The tragedy playing out underwater is much worse, but invisible to most," Fears wrote in the *Washington Post* story that won a 2020 Pulitzer Prize. By 2019, when Fears visited Dillon in Tasmania, more than 95 percent of the island's giant kelp forests had disappeared in the preceding few decades.[10] As the kelp goes, entire communities of underwater life go with it, and new warmer-water species move in. Urchins especially boom and graze the stubble off the rock, preventing kelp from growing back and ushering in ecosystem change.

We humans like to use trees as symbols of time: old, slow, deep, eternal. Perhaps we need to extend the metaphor underwater to the kelp and to picture these marine heat waves as a coastal inferno tearing rapidly through the world's submarine forests. "Despite their importance . . . there is a relatively low level of public awareness about seaweed forests compared to ecosystems on land or coral reefs in tropical latitudes," writes ecologist Adriana Vergés.[11] "Because these underwater forests are 'out of sight, out of mind,' their value and disappearance has largely gone unnoticed by the public."

Kelp can grow up to two feet per day, more quickly than any other seaweed. It also goes through major boom-and-bust cycles. In Southern California, for example, once common kelp forests nearly vanished in the 1950s and 1960s. One bed off San Diego had "yielded 75,000 tons of kelp a year" in World War I, marine biologist Wheeler North wrote in 1972, and then over the subsequent decades nearly disappeared.[12] A strong El Niño had weakened the forests, but the bigger culprit, North declared, was sea urchins: "Like locust plagues on land, hordes of these bottom grazers move through the forests devouring the lower parts of plants." The "plague" mirrored a similar population explosion in coral-eating sea stars that had decimated tropical corals.[13] North concluded that the urchins were thriving in the absence of their traditional predator the sea otter and showed that, in one area where sea otters had been protected, the kelp still thrived. Following North's work, people instituted urchin control and kelp restoration programs, and cleaned up sewage outflows—and by the early 1980s, the Southern California kelp forests had recovered.[14] Kelp, in other words, is part of a complex tangle of marine life, and it's not in any way easy to determine quickly how or whether or why it's declining, and just how serious or long term a decline over a few years or even decades might be. Nonetheless, by the late 2000s, scientists had become increasingly concerned about the potential for global kelp declines and the ability to bounce back from repeated crises.

Just as meadows and grasslands spring up around and between the trees on land, the great undersea forests have their meadows, too, often in the form of vast beds of underwater flowering green plants called seagrasses. Like terrestrial meadows, seagrass meadows grow in open spaces with abundant sunshine, typically in bays and estuaries where waters are calmer and shallow. They are home to innumerable small and colorful animals, and they are nurseries for many of the world's most important commercial fish. Unlike kelps, algae that look like plants, seagrass is a true plant that evolved on land and moved, many millions of years ago, back into the sea.[15] It has a broader temperature tolerance than kelp, and many seagrass species thrive in the tropics, but as photosynthesizing plants, all seagrasses are extremely sensitive to light and water clarity.[16] Over the hundred million years since the seagrasses split from their terrestrial cousins to return to the sea, flowering plants on land have proliferated and become the dominant form of plant life on Earth, with something like 300,000 known species.[17] As one small proof that it is exceedingly difficult for land-based life to rejoin the ocean, there are only around seventy known species of seagrass.[18]

In the same way that, on land, more people seem to enjoy looking at trees than looking at grasses, seagrass is perhaps less iconic than kelp. Despite their differences, seagrass and seaweeds are linked in both the scientific literature and in human culture. They often grow together and face the same threats, and their health tends to rise and fall in parallel. Seagrass, like kelp, faces threats from warming, stronger storms, and invasive species. Coastal urbanization and development affect both particularly. Seagrasses need clear, clean water, and urban runoff carries into the water all the fertilizer and nutrients that people apply to their terrestrial environment, runoff that feeds algae that coat the surface of the seagrass and slowly suffocate it. When they decline, both seagrass and kelp tend to be replaced by fast-growing opportunistic algae.[19]

Seagrass, like kelp, benefits people enormously. By one scientific estimate, more than one-fifth of the world's most frequently caught fish use seagrass as nursery habitat.[20] Rooted deep in the nearshore sediment, seagrass stores massive amounts of carbon—up to 10 percent of the total ocean's carbon sequestration, while it covers only about 0.2 percent of the global ocean.[21] Like trees on land, seagrasses take in carbon dioxide and produce oxygen, benefitting all the oxygen-breathing animals nearby. By soaking in carbon, they also can reduce the effects of ocean acidification, to the extent that some aquaculturists have wondered if the seagrass growing near their shellfish beds acts as a buffer against the changing oceans. Sea otters, the quintessential kelp forest keystone species, also occupy seagrass beds, and in areas where otters are present, eelgrass genetic diversity rises and the meadows are healthier overall.[22] The mechanism that researchers propose has all the beautiful intricacy of ecological relationships: the otters eat crabs, which eat sea slugs, which eat algae, which suffocate seagrass. Add otters and you reduce crabs, increase sea slugs, reduce algae, and benefit the seagrass.

Years before the first global analysis of seagrass extent, most scientists expected that the plant was in decline worldwide.[23] Sea otters were hunted nearly to the point of extinction in the North Pacific, coastal development has intensified, and climate warming has accelerated—all spelling bad news for seagrass meadows. In 2009, a team of international scientists finally put numbers to it, pulling together all the available data and concluding that seagrass meadows had declined "in all areas of the globe where quantitative data are available."[24] Between 1879 and 2006, the researchers wrote in their 2009 paper published in the *Proceedings of the National Academy of Sciences*, about 30 percent of seagrass had disappeared in the sites for which they had direct observations. Worldwide, they estimated, as many as 51,000 square kilometers of seagrass—an area one and a half times the size of the Serengeti—might have been lost.

If the meadows had started to disappear, what about the forests? When the global seagrass paper came out in 2009, kelp researchers wondered if they could document the same history. A group of thirty-seven international scientists led by Kira Krumhansl, a marine biologist at Fisheries and Ocean Canada's Bedford Institute of Oceanography, started to gather data from all the different regions of the world where kelp grows. Like the seagrass researchers, Krumhansl and her colleagues expected they might find shrinking kelp forests, so they especially wanted to move beyond a simple finding of decline to map the pattern and distribution of change. Two possibilities stood out, each suggesting a different path to recovery. If kelp showed a nearly uniform decline everywhere it lived, like coral, it might mean that the primary source of stress on kelp was global change. On the other hand, if the pattern showed kelp decreasing globally on average but at widely different rates depending on where it grew, it might mean that local effects might play as much of a role as global change—which could suggest in turn that local restoration might help avoid the worst effects of warming waters.

Having gathered data from thousands of different sites on every continent where kelp grows, the scientists reported in 2016 a modest overall decline—but one dwarfed by regional variability. Kelp is "strikingly different" than seagrass, or coral, or fish, Krumhansl and her coauthors wrote.[25] In about one-third of the places they worked, the scientists had found major kelp declines since the 1950s. In another third, there seemed to be no difference over the decades. In the remaining places, including Southern California, where kelp had bounced back from its lows in the 1970s, kelp forests had actually increased.

Krumhansl grew up exploring the rocky tide pools of New England. She studied marine biology as an undergraduate and, like many marine biology undergraduates, had gone to school hoping to turn her love of the ocean into work as a tropical reef ecologist or perhaps a tracker of whales and dolphins. She soon found, however, that tracking pelagic marine animals across the open ocean could be frustrating. She rarely ever saw the thing she was studying and spent most of her time looking for it, and she realized she liked being able to see the things she worked with. A tide pool ecologist she worked with suggested to her the value of rooted study subjects. "He was like, 'Barnacles are great!'" Krumhansl told us. "'They're everywhere, you can walk in anywhere and see them!'"

Krumhansl visited Dalhousie University in Nova Scotia and found an adviser who had studied the interactions between kelp and urchins. She liked the idea of following something over the long term and, by coincidence, that was one of the questions scientists around the world had about kelp as the effects of climate change became more and more apparent. "What's going on in Australia, that was one of the turning points for how people thought about kelp forests," she says. "Those pulse disturbances, heat waves or storms in combination with a warming trend, are what cause those massive declines."

No matter where in the world it grows, kelp needs the right water temperature. Extreme heat waves like the one that struck Southern Australia devastate kelp forests anywhere. But for Krumhansl, the pattern that emerged from the international research effort suggested that local conditions could be improved to help kelp forests weather some of those global changes. In Nova Scotia, where she now works, kelp might have survived temperature shocks if not for invasive species. An introduced bryozoan (a colonial invertebrate, sometimes called "moss animal" because of its tendency to live on top of other habitats) crusts over the surface of kelp blades, preventing them from absorbing nutrients and turning the kelp brittle so that it washes away in storms.

Another challenge that Krumhansl encountered while on the East Coast, however, is simply getting people to notice. The kelp forests of the Atlantic often don't extend their canopies to the surface and so, in many cases, they lie just out of sight and thus out of mind. "There's a pretty significant and persistent decline in kelp, widespread in Nova Scotia," Krumhansl says. "But there isn't this large awareness about it."

Up until the late 1990s and the publication of papers about tide pool species on the move or about coral research following the 1997–1998 El Niño, a marine scientist might have reasonably worked without a particular concern for global change affecting their research site. This does not mean that there were no concerns. Sylvia Earle, writing about the kelp forests of Catalina Island in *National Geographic* in 1980, wondered about the effects of California's rapid development following World War II and started researching kelp on Catalina Island, twenty miles offshore of Los Angeles as a way to study a more protected forest. Still, her research then was largely descriptive: what lived where, what ate what, how long kelp lived, how fast kelp grew. Earle's conclusion calls for more attention to pollution: "Knowing that millions of years' worth of revolving seasons—of life and death and life anew—are so readily and instantly vulnerable to our unconscious actions, should we not take notice?"[26]

Krumhansl did not train as a scientist in a world in which kelp forests were static, long-term features. The coastal waters off the Canadian East Coast have warmed rapidly over the last few decades. By the time Krumhansl started surveying kelp there, it was already in decline. On land, deforestation cascades across the landscape, causing changes that range from plant and animal life to the way water moves through the ecosystem. When a kelp forest disappears, it leads to a similar ripple effect, affecting not just marine life but the basic physical structure of that region of the ocean.[27] Like many other younger scientists we interviewed, Krumhansl talked about not having the luxury of simply observing and taking notice of how the ocean or ecosystems worked, unlike previous generations of scientists. And the Canadian kelp shows that even ecosystems out of sight cannot be out of mind. "Even since I've started working there's been large changes," she says. "Even places where I would have worked as a PhD student in Nova Scotia now look totally different. These relatively short-term changes that have occurred are quite dramatic."

Scientists used to say that early humans walked into North America across a land bridge over the Bering Sea. The land bridge is still visible just below the surface of today's ocean, in the barely inundated coastal plain surrounding the long, arcing chain of islands from the Kamchatka Peninsula in Russia to the Aleutian Islands and the western edge of Alaska. In the last decade, however, scientists have started to rethink the bridge metaphor. One of the challenges is how rapidly people seem to have spread across the continent. Although the land bridge was above water by 35,000 years ago, it was exposed because so much of the world's water was held in massive glaciers covering North America.[28] Those ice sheets formed a nearly unbroken wall that would have prevented people walking down the continent itself. It certainly would have slowed them enough that it is unlikely they could have been able to reside 8,000 miles away in the southernmost parts of Chile by 14,000 years ago, the fairly widely accepted date for Monte Verde, South America's oldest archaeological site.[29] Researchers working at Monte Verde have also found nine different species of ancient dried seaweed, also dated 13,000–14,000 years old, showing that those early people used coastal resources and traveled widely enough to harvest them outside their own area.[30]

It has become clearer, many scientists now say, that in addition to *walking* into the Americas, people likely *boated* in, following kelp forests and salmon runs across Beringia and down the North American West Coast even while the mainland ice was still in retreat. This fits with other ways of understanding human origins in the New World. North American Indigenous creation stories often begin with a water world, establishing their long histories as ocean people.[31] "As the coast warmed, starting around 17,000 years ago, and Cordilleran ice melted from the mountainous coastlines," ethnohistorians Nancy Turner and Darcy Mathews write, "seafaring people could have lived within and traversed the Pacific Rim entirely at sea level, without major obstructions, and within an ecosystem providing terrestrial and marine resources." Archaeologist Jon Erlandson called this the "kelp highway hypothesis" in an influential 2007 paper.[32] "Kelp," in this case, becomes a stand-in symbol for an entire marine region, a distinct yet universal place between the tide pools and the offshore depths where people on every temperate shoreline on Earth find themselves at home.

As a graduate student at Oregon State University in the late 1980s, Karina Nielsen jackhammered eighty-four shallow basins into a rocky, wave-washed shelf on the Oregon Coast. With full control over exactly what happened in each of her artificial tide pools, she could fine-tune one part of the system to see how it would ripple out to everything else. Then, for the next three years, she watched. Over the years, patterns emerged slowly, not just in the creatures in her tide pools but in the ocean itself.

"I knew that site, the rhythm of that site, for those three years, incredibly well," Nielsen says. "I developed an impression of seasonality, a little bit of year-to-year variability, from being really really intimate with one place, for a long time. I felt like I started to get to know the system. How systems work." She told us that she felt she had learned a lesson about variability that would help her understand future changes.

In 2014, changes started to arrive in a manner wider and more extreme than Nielsen had been able to imagine for the West Coast system she thought she knew.

Under the grip of an extraordinary weather pattern that led to the American West's worst drought in a millennium (and the infamous 2014 polar vortex on the North American East Coast), the entire eastern North Pacific Ocean experienced a marine heat wave. In some places, sustained temperature anomalies, the difference between the measured temperature and a long-term historical average, reached 7°F (3°C).[33] Scientists looking at temperature anomaly maps saw half the ocean basin painted in red, and nearly all of it above average, by the end of 2014. Mosaic patterns of warmth and cold aren't unusual in the ocean, but the persistence and extent of this event stunned oceanographers and ecologists. Marine heat waves had been rare in the oceans before the mid-twentieth century, researchers found, by using past sea-surface temperature records, with only 2 percent of the world's oceans reaching extreme heat—defined as temperatures above the 98th percentile in the preindustrial era—in the years before 1920. The number slowly climbs, however, until it crossed a threshold in 2014, when more than 50 percent of the entire ocean hit temperature extremes.[34] At its 2014 peak, warmer-than-average water covered an area in the North Pacific nearly the size of the continental United States, and it extended to depths of more than 300 feet.[35] One atmospheric scientist nicknamed the massive pool of hot water seemingly spilling out of the Gulf of Alaska "The Blob."[36]

The ripple effects of the heat wave touched the entire West Coast. As the warm water lapped against the North American continent, coastal temperatures rose on land, especially at night when ocean breezes ordinarily cool major cities like Los Angeles and San Francisco. Warm-water plankton marched north, displacing the higher-calorie cold-water plankton preferred by fish, bird, and marine mammal predators.[37] Soon, the top of the food chain started to feel the effects. Dead seabirds washed up on Northern California beaches by the thousands. Starving seals and sea lions crawled onshore; one badly malnourished, eight-month-old sea lion pup made headlines in 2016 for seeking refuge in the booth of a waterfront restaurant in San Diego.[38] Toxic domoic acid outbreaks, fueled by warm-water currents bringing new species of phytoplankton into coastal areas, poisoned sea otters and shellfish, leading to the closure of commercial crab fisheries and further economic devastation for coastal towns.[39] And kelp forests, for so many the symbol of the West Coast's interconnected nearshore ecosystem, started to disappear. Between 2014 and 2016, 90 percent of Northern California's bull kelp forests vanished.[40] In 2017, Nielsen watched abalone that normally live in deeper water crawling into the tide pools, looking for algae to eat as they slowly starved without the kelp. It was, she told us, "the strangest thing I've ever seen," the marker of a system so completely upended that past studies offered little context to help understand it. "We are seeing things that no one else has seen before," Nielsen says. "The changes are really big. They seem to be invisible to everyone else. What would you notice? Unless you saw something in the newspaper. It's this traumatic change that's invisible."

Another big change was happening in the kelp forest ecosystem, and its ripple effects have now affected the entire coastline from Alaska to Baja California. In the summer of 2013, long-term monitoring sites on the Washington coast started reporting an illness in sea stars. Within a few months, similar symptoms had been seen in southeast Alaska and around Vancouver. By October, reports had come in from Orange County in Southern California, and sea stars had died in an aquarium in the San Francisco office of the Gulf of the Farallones National Marine Sanctuary.[41] The stars would get sick with a wasting disease, their arms would start to fall off, and within a few days they would seemingly melt away. Marine disease expert Drew Harvell recalls hearing about it over the summer but thinking the outbreak would pass quickly. In December, however, the wasting sickness reached Seattle, where she was working at the University of Washington's Friday Harbor Laboratory. Harvell and several researchers decided to do a survey at Alki Beach, a popular sand strip at the west edge of Seattle. "It was unbelievable," she says. "Awful. These arms crawling around by themselves."

Harvell and colleagues have spent nearly a decade trying to understand exactly what happened to the sea stars, and the picture is still confusing. At first, the die-offs appeared to be caused by a virus, but more work showed that there wasn't a strong relationship between the virus and disease symptoms except in one particular species—and in fact, the virus that looked like the culprit at first might be a normal part of the sea star microbiome.[42] Some studies showed a link between warmer water and wasting, while in Oregon, researchers found sea stars wasting away more often in colder than normal water.[43] They died in public aquariums, where the disease clearly spread from one star to the next. At the end of 2021, researchers suggested that perhaps the die-off happened as a result of a change in the microbes that live between the sea stars and the water, which then resulted in oxygen deprivation and the sea stars suffocating.[44] Whatever the cause, they died. Between 2013 and 2016, millions of individual sea stars melted away and decayed on beaches up and down the West Coast. Some species in some places were wiped out entirely. Some, like the ochre stars, have since made a comeback, but the sunflower star has now been listed on the International Union for Conservation of Nature's (IUCN) "red list," indicating that the species is critically endangered. Scientists celebrate every discovery of a sunflower star in the wild and are working to convince them to reproduce in the lab. "We are now running the world's only captive breeding program for the world's only endangered sea star," Friday Harbor Marine Laboratory scientist Jason Hodin told a *Reuters* reporter in 2023.[45]

Bristling with arms, sunflower stars are among the largest sea stars in the world and can grow to the diameter of a car tire. Fast-moving, solitary hunters, they are voracious predators of sea urchins. Remove 80 to 90 percent of them from California to Washington and inevitably ecological repercussions follow.[46] As the warm water flowed through California kelp forests, kelp-eating urchins boomed. Kelp strands that anchored and started to grow were devoured. With the kelp chewed to stubs and unable to regrow, only the rock and the carpet of urchins remained, a condition known as an urchin barren. In Northern California in 2020, with 90 percent of the bull kelp gone, the sea stars still absent, and urchin barrens still proliferating, nonprofit organizations and the state of California started paying scuba divers to kill as many urchins as they could, by collecting them with rakes, smashing them with hammers, or sucking them up into vacuums.[47]

They had reason to believe it would work. Kelp restoration has been tried around the world and frequently involves humans removing urchins. In Japan in the early 1930s, fishers tried to protect valuable kombu kelp with urchin culls.[48] Off the coast of Palos Verdes in Southern California in the 2010s, scuba divers spent nearly 6,000 hours on an urchin removal project, smashing an estimated 3.4 million urchins over four years. Kelp canopy increased in the restoration sites, as did the number of fish living in the kelp. The comeback stalled, however, in a year when warm El Niño water flooded the coast.[49] Around the same time, at Porsanger Fjord in Norway, local authorities spent two years attacking urchins with quicklime, which dissolves the urchin's hard outer skeletons. After two years and enough quicklime to destroy urchins over 270 acres of seafloor, kelp coverage increased.[50] In Korea, as part of a major and ongoing project to restore declining kelp forests, researchers brought 200 urchins into the lab to test the urchins' response to a variety of electrical signals, with the idea that they might develop an electrical tool to repel them.[51]

Ecological relationships are complicated. In a world of global change, even scientifically informed human efforts at protection or restoration can fail. The converse is true as well. Sometimes, in an ongoing ecological catastrophe, we just get lucky. In 2020 and 2021, a strong upwelling season cooled the water off the Northern California coast. The sunflower stars had not come back, the urchins had not died off, but the bull kelp resurged. By the end of 2021, it had doubled in size from the year before.[52] The lesson, conservationists tended to say, was that we cannot plan on a stable or predictable big picture in a changing world. But we can know our local environment and give the kelp the best chance we can give it to survive the warming world.

Across nine main species of kelp and thousands of miles of the Australian coastline, almost all the kelp is in decline—and, as elsewhere, this is the result generally of a combination of local and global, physical and biological conditions. Paralleling California, the overfishing of an urchin predator—in this case, the spiny lobster—led to urchin barrens in the strait separating Australia and Tasmania.[53]

Kelp forests wrap around the Australian continent where the water is cold enough to support them. Head south along the east coast from the tropical colors of One Tree Island and the Great Barrier Reef, and the shallow water transitions from coral habitat to kelp habitat. Kelp inhabits the harbor beyond the Sydney Opera House, encircles the rocky coasts of Tasmania, and shapes vast reefs of its own all along the massive coastal indentation known as the Great Australian Bight, the geologic marker of Australia's long-ago undocking from Antarctica. Australian kelp comes in a variety of species, from the sequoia-like giant kelp—the same species found in Southern California, one of the modern kelps that has crossed the Equator—to spreading underwater shrublands of brown algae called fucoids.

Warm waters, expanding southward in the same heat waves that bleached the Great Barrier Reef, decimated giant kelp forests around Tasmania. In Western Australia, extreme marine heat weakened the dominant common kelp, and warm-water-loving, kelp-eating fish moved in to cause further harm. Along the south coast and around metropolitan areas like Melbourne and Adelaide, urchins overgrazed on kelp, and pollution and urban runoff allowed turf-forming algae to bloom in carpets that prevented the kelp from growing back. In metropolitan Sydney, a kind of kelp called crayweed once grew across seventy kilometers of shoreline and in the 1970s and 1980s just up and disappeared. No one noticed exactly when it went, but it's there in historical photos and herbarium specimens, and in 2008, Australian researcher Melinda Coleman published a paper showing that it was gone.[54] Although Coleman couldn't say for sure what had killed the crayweed, its disappearance coincided with several decades of sewage outflow into the harbor. What intrigued Coleman and many other Australian scientists was that the sewage outflow had been stopped. Water quality in Sydney Harbor had improved dramatically through the 1990s and early 2000s. The rocky barrens where the crayweed once lived were suitable for a comeback. The kelp, however, hadn't returned. The gap in kelp coverage in water once again clean enough to support it suggested possibilities for restoration.

A kelp forest can be recovered in two basic ways. One is to remove other sources of immediate stress and hopefully allow the kelp to grow back naturally. Around Sydney Harbor, however, cleaning up the sewage had not resulted in a return of crayweed. The other method is to plant kelp directly. But as any home gardener knows, it's not as easy as it sounds to grow something you really want to grow. You have to deal with pests, light, nutrients, and environmental conditions, only some of which you can control. Moving the entire operation underwater adds even more complexity and cost, which is why direct kelp transplant projects often fail.

In 2011, Australian scientists dove into the water off Sydney with a small batch of crayweed colonists. The kelp was zip-tied to plastic mesh, which was then bolted into rocky areas that the historical record showed once sustained a thriving forest. To the scientists' delight, the transplant worked. A year later, the crayweed was reproducing on its own, and what the scientists quickly labeled "craybies" began to appear. In a second, slightly larger operation in 2012, the researchers planted a larger area with crayweed, and although it died back the following year, the crayweed babies grew and thrived, eventually expanding a few hundred meters in every direction from the original planting site. This raised a somewhat surprising challenge: How, in the modern media environment, to spread the good news? In a survey, the researchers found that most members of the public didn't know the crayweed had ever been there, didn't notice its disappearance, and incorrectly assumed that coral, not kelp, had been the dominant ecosystem off Sydney. It was more surprising to learn that many people didn't know that water quality had improved in Sydney since the era of uncontrolled sewage outflows. The majority of the 600 people surveyed believed water quality had become worse.[55]

To take the operation beyond a few successful experimental plots, the researchers needed more funding and public support. They came up with a science communication plan, a collaboration with artists, and a crowdfunding request that quickly went viral: plant an underwater tree for Christmas. In four days, they reached their fundraising goal. In two months, they doubled it. The public support led to institutional support, and Operation Crayweed now has expanded to eleven sites around Sydney, including, in 2017, just offshore from the iconic Bondi Beach. Even in an era of global bad news, when people agree to take care of the places they live, it can yield startling improvement. "Researchers are increasingly recognising that sustainable management of our coastlines requires a better understanding and appreciation of the emotional bond that people have with marine systems," marine scientist Adriana Vergés, one of the scientific leaders of Operation Crayweed, and a number of colleagues wrote in 2020.[56]

What can we make of a locally successful project in a globally distressed world? The Australian researchers argue that any scientific restoration project should involve the people in whose backyard it takes place. But Operation Crayweed had a unique twist in that it was restoring something that had already disappeared. The crisis had come, and passed, and the operation was to heal the damage. Halfway around the world, a Chilean scientist who'd earned his degree studying the decimated kelp forests of Tasmania asked another question: Could you engage the public in conservation measures on behalf of kelp *before* the crisis becomes apparent?

Alejandro Pérez Matus grew up in Santiago, Chile, about sixty miles from the ocean. He swam competitively through high school, but always in pools. His junior year in high school, his then-brother-in-law, an actor, was cast in a Chilean soap opera to play a scuba diver. Pérez Matus met his brother-in-law at the pool one day, along with the instructor who was teaching him how to scuba dive. The instructor offered Pérez Matus a few lessons, too. He liked them so much that at sixteen years old, he had found a calling.

At the time, he recalled to us, the Chilean coast did not have a diving industry. There were no training centers or equipment stores. He worked several jobs to buy his gear, and then he just went to the ocean on his own. Friends up and down the coast helped him explore: the cold fjords off Southern Chile and Patagonia, where the giant kelp *Macrocystis* grows thick and marine giants feed; Central Chile, clear and cathedral-like and dominated by thinner strands of *Lessonia*; the northern desert coast, beloved by surfers, where a diver can submerge from the parched and plantless dust of the Atacama into a sudden riot of life and color even in the shallows.

Pérez Matus went to college in the United States, where he explored the North Pacific. He returned to Chile for a master's degree on the desert coast in Coquimbo and then moved to the University of Wellington for a PhD program in kelp forest ecology. In 2008, as part of his dissertation, Pérez Matus surveyed the kelp forests of Tasmania. Now, he says, "all the kelp I surveyed there is gone."

Chile has not faced the dramatic warming events that Australia has. In fact, ocean temperatures have cooled off the coast of western South America. But kelp in Chile faces a separate challenge that the mostly protected Australian kelps don't: harvesting. Humans use an awful lot of kelp.

Like many other forms of brown algae, kelp constructs its cell walls in part from a polymer called alginate.[57] From a human perspective, this alginate is something of a miracle substance. Extracted from dried kelp and combined with calcium chloride to make a salt, it can be used as a gel and thickener. Kelp-derived alginate appears in toothpaste, pharmaceuticals, ice cream, and pudding, and as an additive in some dairy products.[58] It is almost impossible to avoid using kelp products in everyday life, and if you have used kelp products, you've likely used Chilean kelp. Chile supplies about 40 percent of the global alginate industry: hundreds of thousands of tons of kelp every year.[59]

When Pérez Matus first started college, most kelp was collected from the shore. But over the last two decades, fishers have increasingly specialized in harvesting it live from the seabed. A quota system covers around seventy small bays and coves, allowing up to three acres of kelp harvest per cove. The effect of this is still hard to assess. Pérez Matus runs one of the first labs dedicated to studying Chilean kelp forest health. "We have to take action," he says. "And without information it's hard to take good action."

In 2016, Pérez Matus joined kelp researchers from around the world to work on the global assessment led by Canadian scientist Kira Krumhansl. The paper identifies Chile as one of the fastest-declining kelp forests in the world, showing, the authors wrote, just how important local action is. Because the water has remained cooler, kelp could potentially bounce back much more quickly from overharvest—but not entirely. Although small coastal strips of the Northern Chilean coast haven't warmed as much as many other places, climate change is already transforming Chile.[60] Land temperatures have increased, and the country's internationally renowned wine industry is moving south, following the cool temperatures and rain that it needs. There are plans for major new highways and pipelines to connect north and south and open Patagonian resources to the rest of the country. At some point, change will reach the forests offshore as well. Pérez Matus recently dove in southern giant kelp forests described by Charles Darwin in the 1830s and closely studied again by marine biologists in the 1970s. After witnessing firsthand the decimated giant kelp forests in Australia, he said, it was like going back in time. "So then imagine diving [in Patagonia], it's great," he said. "It's like seeing the past."

The lessons of Australia and California still apply. In an unceasingly warming world, how long can coastal Patagonia look like it always has? If the kelp forests of Patagonia, an inspiration unchanged for 200 years, can't be protected by their remoteness or by law, what other options are there? Pérez Matus told us that his goal is to connect Chileans to the ocean and to learn more and teach more to try to change the relationship people have to nature. But even as he learns, he watches the world change. Surf breaks he once enjoyed have been wiped out by changing sand deposits at the beach and the impacts of mining. Places where he used to see fish he now describes as "empty." Plastic, still rare even a few decades ago, is everywhere. The future seems uncertain. "Without changing our perspective of our economy based on natural resources," Pérez Matus said, "it's going to be very difficult to maintain and to live with that uncertainty."

Haida Gwaii is a large archipelago fifty miles off the northwest Canadian coast, an eight-hour ferry ride from Prince Rupert, the nearest mainland city. Its bowed western edge lies exposed atop the precipice of the continental shelf, and wind, rain, and massive waves batter the coast. The southern tip of the islands might be one of the closest places to the ocean's abyssal plain that you can stand and still be on the North American continent—ten miles out from the rocky beaches of Kunghit Island, you would be past the steep continental slope and floating over 6,000 feet of water. Strong offshore upwelling and currents bring incredible marine biodiversity, from whales to nesting seabirds, to fish and shellfish, to vast kelp forests and seagrass meadows. The islands have suffered ecologically since colonization, but still some call them the "Galapagos of the North." "It's quite wounded now," says Kii'iljuus Barbara Wilson, a Haida elder who has lived on the islands her entire life. "But when I was a little girl there were all kinds of flowers, and birds, and medicinal plants and food plants. And the intertidal too was very prolific."

The roughly 200 islands of Haida Gwaii are the traditional home of the Haida Nation, and about half of the archipelago's 5,000 residents are Haida.[61] Wilson lives in Skidegate, a village of a little more than 800 people on a sheltered, east-facing inlet. Her house sits about twenty feet above the high-tide line, and runs onto a grassy lawn and then a gravel beach. On the sunny day we talked to her on a video call, Wilson showed us the view out the window, then picked up her phone and led a tour down to the water's edge, where ankle-high waves rolled on the rocks and a line of seagulls stood watching. "I'm a grandmother, I live by myself, and I love what we're doing," Wilson told us, by way of introduction.

Wilson was sent to one of Canada's residential schools as a teenager. Her grandmothers, traditional Haida knowledge holders, died when she was young. Her mother attended European-style schools, married at fifteen, and never had a chance to learn many of the ways that had once supported 30,000 people on Haida Gwaii. In the 1960s, when she was in her twenties, Wilson started to do her own research, "trying to find out who I am, who I was," she says. "I ended up a lot of times teaching my mother different things."

Her father was a hunter, trapper, and logger, and knew the land and coast, but like Wilson's mother had grown up at the height of Canadian colonial suppression of Indigenous knowledge. He rarely spoke of what he knew until he was much older, when, in his nineties, he would eat dinner with Wilson and her brothers and then drive them around the islands, telling them stories. Wilson remembers that, when she asked how he knew things, he would always say, "So-and-so told me," reflecting the complex flow of information and knowledge in the Haida oral tradition.

In 2019, at age seventy-six, Wilson graduated with a master of arts degree from Simon Fraser University and became the first person to defend a thesis in her hometown on Haida Gwaii. Her project examined the ways the Haida have been affected by and can respond to climate change, and the legacies of colonization that have worsened its effects and made responding more difficult.[62] The thesis opens with an often-cited quote by climate scientist Ken Caldeira comparing carbon emissions to the mugging of little old ladies—to argue that the goal should not be to set acceptable targets but to reduce emissions entirely to zero. It ends with a note about the way her ancestors once tracked time by the world around them: kelp growing, petals dropping from the salmonberry plants and the berries changing color, the particular tinge of blue on an early-season halibut's flesh, the salmon traveling past the island's western edge on the way south to Oregon or California. "There was this knowing that was deeper than just the surface of the ocean, that was associated with the land," Wilson says. "Those kinds of things are the kinds of things everybody is missing when they don't use traditional knowledge, the old knowledge that our people developed from living for thousands of years on the land. Thousands and thousands."

Between 2005 and 2011, ethnographers and Haida elders worked together to assemble a three-volume report on Haida traditional marine knowledge.[63] It documents connections to plants and animals, seasons, and places. The Haida made use of about forty marine species in the vicinity of Skidegate alone, it says, some seasonal and some year-round. They fished, foraged, hunted, gathered, and trapped. Now, Wilson's thesis notes, climate change is remaking the world so quickly that even the thousands of years of Haida knowledge, encompassing many healthy ways to respond to natural boom-and-bust cycles, has been pushed to its limit. "We have adapted as we can," Wilson wrote. "Now we are a place where, even though we are resilient, it may not be enough. As ocean people our food sources have been impacted especially in *Xaana Kaahlii* [Skidegate Inlet] where we live."

Until colonization, kelp forests and seagrass meadows surrounded Haida Gwaii, providing a rich variety of food sources. That started to change when European fur traders arrived in the 1700s to hunt sea otters. As the price of otter pelts increased, Russian, American, and Indigenous hunters rapidly drove down the population all along the North American West Coast,[64] and otters had disappeared completely from the Haida Gwaii archipelago by the early 1900s.[65] As the otters died, sea urchin populations boomed and started to overrun the kelp forests. An intensive commercial fishery started, rapidly decimating abalone and herring populations. Invasive tunicates linked to warmer waters started to arrive.

In the 1970s, the government of Canada decided to reintroduce sea otters to Haida Gwaii, partly as a way to restore and protect the health of the kelp forests. But now the otters were returning to an impoverished coast. Voracious predators that are commonly described as eating over 20 percent of their body weight in a single day, otters have become direct competitors with people who rely on the archipelago's remaining shellfish. Once the Haida hunted otters for food and fur and to keep the kelp forest healthy without overexploiting them. Now the otters eat their way through important abalone resources, and federal laws often mean both the otters and the abalone are designated as protected species—effectively preventing the Haida and Indigenous people, from Alaska to California, from using the thousands of years of experience they have in balancing otters, shellfish, and kelp.

"Indigenous people are feeling like they have very little control over a predator that's directly, and very quickly, reducing the numbers and size of many of the shellfish that are a major source of food," Simon Fraser University (SFU) marine ecologist Anne Salomon told *MacLean's* magazine in October 2020.[66] In 2013, Salomon started a project called Coastal Voices to try to bring Indigenous and Western science together to offer strategies for managing the otters.[67] Wilson, who wrote her thesis in Salomon's lab at SFU, joined the partnership as a cultural adviser. "We've got to do something different," Wilson said. "This idea of *Terra Novis* and doctrine of discovery and the idea people could come here and take things without thinking about the consequences for people who lived there, you know, it's quite disheartening at times."

Western conservation tends to draw lines at the water's edge, and on the water's surface itself, a reflection of a way of thinking that separates people from the ocean. Western scientists tend to describe themselves as working in the marine *or* terrestrial realm, or as working at the interface *between* marine and terrestrial, even though decades of scientific research has documented the close connectivity and relationships between these realms. In contrast, in an 1855 treaty signed with the settler governor of Washington State, the Makah people of the Olympic Peninsula demanded territorial rights that extended out to sea, with one Makah chief famously telling the governor, "I want the sea. That is my country."[68] Along the same lines, Native Hawaiians divided their islands into mountain-to-sea communities called *Ahupua'a*, which often connected terrestrial agriculture to thriving marine fishponds.

The marine conservation ideal remains the marine protected area, which mirrors the concept of parks on land: a zone in which fishing and other extraction for human use is banned or at least partially restricted. Over many decades, California, Australia, and New Zealand set up some of the world's most innovative marine protected area networks. In some cases, the development of these protected areas involved an extensive and at times contentious community engagement process, which included scientists, managers, commercial and recreational fishers, and tribes, in utilizing science and human connections to the ocean to map and protect portions of coastal resources.[69] Then the marine heat wave arrived in California, urchins landed in New Zealand, and repeated bleaching events swamped the Great Barrier Reef. By 2020, many of the gains of the world's vast marine parks were under threat due to warming waters.

As this book goes to press, more and more examples of Indigenous comanagement of marine and terrestrial reserves are emerging—for example, in the United States, Canada, and Australia—where Indigenous science and Western science are used in tandem to try to approach marine management and decisions differently. Wilson argues that doing something different means that, rather than drawing more lines, society as a whole needs to change the relationship it has with the ocean. Kelp forests show how important local action is. "If you learn something you have a responsibility to look after it," Wilson says. "When our people pushed off the beach here, they always made sure that everything in their world was OK. They knew there was a possibility they would not come back and they had to make sure the people they left behind were going to be all right. That's what we have to do. It doesn't matter if it's cancer, or COVID, or famine, or fire. We have to look after each other. That's really important. That extends to the birds, the whales, the fish. God forbid, the worms. The snakes."

CHAPTER 4

THE GARDENS

Now is a time to imagine a world in alignment with our ideals, as we embark on our postindustrial age. We will need to feed and shelter ourselves and one another, but everything else is on the table.

—Hope Jahren, The Story of More

The Kingcome River flows out of a glacier in an ice field of the Pacific Coast Range, winds south through a cottonwood-clad valley and empties into the end of a long, narrow inlet in the fjords of western Canada. The inlet makes its way west into what's called the Broughton Archipelago, where glacier-fed streams and milky green inlets twist around innumerable small islands like roots branching off the western edge of the continent. On land, the ice is never far away, but the ocean abounds with life: kelp forests and seagrass meadows, whales, dolphins, seals, sea lions, otters, salmon, herring, halibut, and dozens of varieties of shellfish. The remote archipelago is the traditional home of the Dzawada'enuxw (Tsawataineuk) and other Kwakwaka'wakw First Nations. It is still nearly entirely roadless, with small fishing ports and Indigenous reservations connected to each other and to the mainland by boat.

In 1933, near the peak of Canada's attempt to suppress the culture of its Indigenous people, a four-year-old boy moved to a small Kwakwaka'wakw Nation village on a bend in the Kingcome River one half mile upstream from the river mouth. The boy's parents and grandparents had been clan chiefs, aristocratic keepers of thousands of years of knowledge, and they and other Kwakwaka'wakw clan chiefs had selected the boy at birth to receive specialized training and uphold their traditional knowledge for future generations, in defiance of the government's cultural suppression. Lookouts posted in the forested hills overlooking the strait warned of police boats, so the boy could be kept from forced attendance at the boarding schools that taught Western culture. They named him Adam Dick, but in time he also would be bestowed the chiefly name Kwaxsistalla, a name held by countless clan chiefs before him.[1]

With his move to Kingcome and the village of Gway'i, Adam's grandparents and other elders started him in an intense formal apprenticeship, says Kim Recalma-Clutesi, his surviving partner. From the many elders who gathered or visited, Adam learned highly specialized language, ecology, and tradition. He learned seasons, weather, and tides. He learned the rules for salmon harvest and what to do in the lean years when the fish didn't arrive. He learned that, by mounding soil in the right places in the island's many estuaries, you could increase the output of important root food plants and that his people had been constructing and tending those root gardens since time immemorial. He learned the right season for harvesting seagrass and what the presence of a seagrass meadow meant for other marine life. He learned the one week each year that was best for seaweed harvest, the month when the northwest bush cranberries produced the most fruit, the time of year to fish for halibut. He learned a complex royal form of the Kwakwaka'wakw language used only as part of ceremonial and spiritual practice, and the songs that encoded his people's thousands of years of knowledge.

Adam spent years learning a way of life that, at the time, the Canadian government declared archaic. The government and Christian churches built residential schools because government policy said First Nations people would never be able to participate in the modern world, and the only hope was assimilation. The government banned potlatch gatherings, and in 1921, agents raided a Kwakwaka'wakw potlatch and confiscated dozens of ceremonial objects.[2] Adam learned his lessons in secret, moving between remote islands in a dugout canoe with only his grandparents and other elders, in part to learn all the places he'd need to know as a clan chief, and in part to stay ahead of the police boats.

In 1967, Canadian writer Margaret Craven wrote a short novel, *I Heard the Owl Call My Name*, about a young vicar who is sent to work in the church in remote Kingcome. In front of a backdrop of towering mountains; deluges of rain; and forests of alder, green spruce, and cedar, the vicar learns from the Kwakwaka'wakw. Craven, who visited Kingcome in 1962, based her Kwakwaka'wakw characters on Adam Dick and his family. According to Recalma-Clutesi, it was Adam who taught Craven that the owl's call was an important omen.

———

Pioneer and many scientific accounts describe the Pacific Northwest coast as a bountiful place of plenty, where First Nations people lived in simple harmony with nature.[3] Recalma-Clutesi told us that, on the contrary, Adam learned from a young age just how much work had gone into making it bountiful and how much work was involved in keeping it so. It took a very long time, Recalma-Clutesi says, before he realized how few people knew what he knew.

But Adam lived long enough to see a different world emerge. As the implications of climate change and habitat destruction grew more obvious, scientists and conservationists started to look for better ways of resource management. Their searches, often conducted over many years, led them to Indigenous people who had successfully managed resources and environments for thousands of years. In his later years, Adam saw some Western-trained scientists realize that Indigenous science held answers to ecological questions their own research could not uncover. Indigenous people had said it all along, but for the first time in centuries, resource managers started to listen and understand: the original people of North America did not survive purely by the richness of the environments they were situated in or by foraging in a bountiful wilderness but by using their knowledge to manage the environment and create the conditions for their own success.

These realizations, over a period of a few decades, came with a bitter irony: many Indigenous people were struggling to make sure that knowledge was not lost. Many of the elders who held it had died, often heartbroken, sometimes, like the Ohlone elder Isabel Meadows, having imparted as much of that knowledge as they could into an ethnographer's notebook. Languages that contained within them thousands of years of understanding of ecological connections were disappearing. Young people who might have learned time-honored methods for thriving in an uncertain environment had instead been sent to assimilationist schools, or they were simply taught in public schools that their people were extinct.

Out on the western edge of British Columbia, Clan Chief Adam Dick grew up and started working as a commercial fisher. He continued to move around, living in Kingcome and in Alert Bay, and then with Recalma-Clutesi in Qualicum, where her brother is the Kwakwaka'wakw clan chief. For many decades, Adam held in his head an understanding of the coastal ecology of the Broughton Archipelago unmatched by anyone alive yet nearly unknown to Western science.

Specific marine knowledge can seem unimportant to the way most people live their lives now. But as rocky shorelines, coral reefs, and kelp forests around the world show, local conditions matter. People with local knowledge and local relationships matter in keeping local ecosystems healthy. The knowledge in the disappearing language and culture of people like the Kwakwaka'wakw, and so many other First Nation and Native American groups, reveals ocean declines that the descendants of settlers cannot see. This is a particularly dramatic example of "shifting baselines"—if so few people knew of what a place once was, how would we be able to identify what has changed and what has been lost? It's as if the extensive knowledge humans have acquired over millennia is a set of fine-tipped nerves, and we are slowly cauterizing those nerve ends even as they broadcast alarm signals back to the center.

Melissa Poe, an environmental anthropologist at the University of Washington, told us a story about communities around Seattle and about those that spend some of their days overlooking the Salish Sea. She described smart, hardworking people who find a connection to the sea and enjoy the daily ocean-to-mountaintop panorama on their regular commute. In some neighborhoods, they can sit and watch the sun setting over the deep blue and breathe a deep sense of place and their marine connection. But beneath the surface, there is more complexity and perhaps less human connection. The Salish Sea has a long history of challenges related to pollution, Indigenous rights, and industry. The shellfish once central to Indigenous foods have suffered, including the Pinto abalone, the first shellfish to be listed as an endangered species for Washington State.[4] The population of killer whales has diminished in tandem with the declines in their food source, salmon, and the increasing challenges of toxic pollution and boat noise in the marine environment.[5] "The mismatch between the outwardly healthy appearance of my home waters and what I know to be the underlying threats to their health helps me understand why it is that many people I talk to don't readily appreciate the seriousness of the situation," marine disease expert Drew Harvell writes in *Ocean Outbreak*.[6] "As long as seascapes remain scenic and we can still buy salmon at the grocery store, there seems to be no cause for alarm about the conditions of the ocean."

The sunsets are as remarkable as they have always been in the Pacific Northwest, but beneath that orange glow and shimmering blue water lies a frayed human cultural connection to the ocean. Natural scientists aren't accustomed to thinking about such cultural ties. But to really know the ocean, you need to listen to the people who know it, Melissa Poe told us. She told us that her work has made clear that "human wellbeing vital signs are coupled with ecological vital signs."

We know shellfish have been important to people since our species first emerged. A connection to the sea and its resources threads throughout human history. Archeological finds suggest that *Homo erectus* foraged for shellfish off the coast of Indonesia 800,000 years ago, and *H. neanderthalensis* fished for clams off Gibraltar 120,000 years ago.[7] Archeological sites dated to about 165,000 years ago near Pinnacle Point, South Africa, show evidence of tide pool foraging—shellfish harvests—extending the time line of our reliance on the sea to the "stone age."[8] Shell middens, the sometimes massive piles of discarded shells left by people, and other times large areas of less concentrated shell materials,[9] are "among the most common, and certainly the most apparent, coastal archaeological site type in the world," Simon Fraser University (SFU) archaeologist Dana Lepofsky wrote in a 2015 article.[10] The evidence from Lepofsky's work and others suggests that, even thousands of years ago, people didn't just walk to a tide pool and pick what was there; they actively worked the coast. Shell middens from the Jomon period in Japan, 8,000 years ago, tend to hold mainly large shells, which some archaeologists infer as a sign that people had a conservation ethic to discourage taking young and still-growing animals.[11] The ancient Romans created artificial oyster beds by walling off freshwater lakes and fed their cultivated shellfish to fatten them.[12] And for thousands of years on the Pacific Coast, from Alaska to

California, people like Kwaxsistalla and his ancestors manipulated and managed shorelines to boost shellfish productivity.[13]

In the 1990s, geomorphologist John Harper surveyed the Johnstone and Queen Charlotte Straits of western Canada from a helicopter as part of a project to map oil spill contingencies. He spotted something no one he knew had ever described before: miles of rock walls emerging on beaches at the low tide. He had no idea what they could be. Harper counted 365 of these rock terraces, totaling nearly ten miles of shoreline. Along with an archaeologist and marine biologist, he went to investigate and found clams living thick around the walls, as well as shell middens nearby. Harper could not explain the origins of what he called "clam terraces," he wrote in a subsequent report for the government of British Columbia.[14] It is possible that sea ice had deposited the rocks, he wrote, but the features seemed too regular, the boulders too uniform in size. Given the number of middens nearby, he concluded that it was "probable that the clam terraces were subjected to some degree of modification by aboriginal shellfish gatherers over the thousands of years of occupation in the region."

Harper spent years trying to explain the origins of the clam terraces. At last, according to a National Geographic documentary called *Ancient Sea Gardens*, he got his break. A local fisher and beachcomber suggested he talk to Adam Dick. Harper called an ethnobotanist friend, who called Kim Recalma-Clutesi. It was Halloween 2002, Recalma-Clutesi told us. She drove from Qualicum down to Victoria to meet Harper and his wife, Mary Morris, also a surveyor. Over dinner, they peppered Recalma-Clutesi with questions about the walls, most of which she couldn't answer. They sent her home with a photo. Later, Adam asked what had happened. She showed him the picture. "Ah," he said, "it's a *loxiwey*."

Recalma-Clutesi didn't know the term. No one did, except Adam, and he, as happened often, didn't realize how little the outside world understood of Kwakwaka'wakw traditions. As she retold the story to us, she explained that, in the middle of the night, he asked Recalma-Clutesi to call Randy Bouchard and Dorothy Kennedy, anthropologists they sometimes worked with. When Bouchard picked up the phone, Adam started to sing. His song described the loxiwey, what it was, what it meant, and how it should be tended. The loxiwey was a clam garden. Kwakwaka'wakw people had been making them for millennia as a way of improving harvests.

Finally, they got off the phone. "Why are they doing this?" Adam asked.

"Well," Recalma-Clutesi said, "this man seems to have discovered something."

"Discovered?"

"Well, found."

And Adam answered, "That's fiddlesticks. It wasn't lost."

He had been making and tending loxiwey since he was a child. Elders had described to him not just how to make them but why. There was no accidental component to it—the structures were specific and intentional, their uses and maintenance described in the traditional songs that Adam knew. The term *loxiwey* itself came from the Kwakwaka'wakw word for "roll," and the elders taught that, when you went to a beach, you always had to roll rocks. This was something Adam could feel, his friends and family say; a beach wasn't quite clean until he'd moved a few rocks around to clear the clam beds and reinforce the retaining walls. It was so obvious to him, so deeply ingrained, that he had trouble understanding what about it so intrigued Harper.

"It's like saying to me, 'Do you put yeast in bread, and how much, and when did you start doing it, and which of your ancestors taught you that?'" Recalma-Clutesi told us.

Recalma-Clutesi arranged a tour with Adam and Harper at a series of clam gardens around Kingcome, where Adam grew up. Harper explored all day, asking questions, measuring, running around on the mudflats. At one point, Recalma-Clutesi recalls, Harper and a colleague stopped over a smaller loxiwey to wonder out loud why some were larger and some smaller. They speculated for several minutes while Adam sat nearby, watching. "Kimmy," he said, "I made that garden as a kid."

Harper and his colleagues published their clam garden findings, and then the *Ancient Sea Gardens* documentary featuring this work came out, bringing clam gardens to the attention of a much wider scientific audience.[15] With new interest came new people. One was Dana Lepofsky, who had started studying ethnobotany, turned to archaeology for her PhD, and arrived at

Simon Fraser University in British Columbia with an interest in the human relationship to the environment. Clam gardens had just exploded into the archaeological scene, and Lepofsky told us that she found it easy to switch from plants to clams.

Lepofsky spent much of the next twenty years investigating clam garden sites on the British Columbia coast. At Quadra Island, at the east end of the Johnstone Strait, home of the Laich-Kwil-Tach, southern Kwakwaka'wakw, and Northern Coast Salish First Nations people, Lepofsky and her colleagues found that 35 percent of the total shoreline had been converted to clam gardens and that some of the walls were 3,500 years old.[16] Lepofsky's work documented the same species in shell middens over time, and the shells don't get smaller unlike harvests that are done through wild foraging; these findings support the idea that clam gardens worked to create a reliable food supply even when fished hard. Shellfish are easy to overexploit, and researchers who study foraging emphasize the perils of overfishing them—but their accessibility also makes them exceptionally good targets for sustained farming.[17] In a sense, they are more like plants than wild game: people eat lots of them at a time, and they can't exactly run away. Throughout history, people all around the world have built cultures on oysters, clams, mussels, and snails without driving their food extinct. The coastal people of the Pacific Northwest had found a technological solution that had worked for millennia to guarantee food security.[18]

"There's something about clam gardens that's amazing," Lepofsky told us. "You can't deny that they're management features. When they first started breaking on the academic scene, people were saying 'Yeah, they're not deliberate.' But they're in your face. You can't deny people did them intentionally with a purpose of maintaining the foreshore."

In a 2015 paper coauthored with Adam Dick and Kim Recalma-Clutesi, Portland State University cultural anthropologist Douglas Deur writes that, by the time archaeologists and marine scientists became interested in the clam gardens, Adam may have been the only person on Earth who had learned about them from practitioners who had been present in the nineteenth century, when the gardens were widely used by the Kwakwaka'wakw.[19] As they have become more popular in the media and scientific literature, and as First Nations groups have been able to reclaim traditional practices, it has become clear to scientists that clam gardens are a widespread feature of the western North American coastline. After a century of neglect and academic suspicion, clam gardens are reclaiming their role as an essential part of the future of the Pacific Northwest.

"I remember trying to get modest pockets of grant funding to support the work we were doing," Deur told us. "People thought it was intriguing, but I got a few comments like, 'This is absurd. People of the North Coast were not capable of these things.' It was very fringe. These days everyone knows it and everyone claims to have been a discoverer."

In that same 2015 paper, Deur and coauthors argue that "over the last 15 years, assessments of previously overlooked practices of plant cultivation . . . and management of fish resources . . . have expanded nearly monolithic representations of Northwest Coast aboriginal peoples as 'Hunter-Gatherers' towards an understanding of these peoples as active resource managers and cultivators." Deur said that, as an early career researcher in the late 1990s, he felt he had to step outside academia to support his work with Adam on clam gardens and traditional ecological knowledge. Now, he said, graduate students "operate in a world where everybody knows this."

Arriving at a Pacific Northwest beach, you may first notice the mixture of sand and cobble, the fleshy green seaweed drying on rocks at low tide, the smell of the salt air. There's the distinct feeling of endless curving inlets and bays; small islands and peninsulas with a background of forest; and, in the distance, often snow-capped mountains. Land and sea merge into a seamless patchwork of water and mountains and forests. But this scene may also be the location of a carefully arranged set of rock walls, delineating productive clam habitat. The gardens resemble terraces made for gardening on hilly land. Boulders and cobbles are arranged to rim the clam harvesting area, to create a wider and flatter home for clams. The human gardeners who kept these clam gardens turned sediment to alter the slope of the beach, harvested selectively to maximize clam productivity, and added ground-up or crushed shells to the sediment to provide a cue for juvenile clams to settle and burrow. The rock walls were built to curve with the shoreline and meet the edge of the water at low tide; between the constructed wall and the often forested edge of the beach is a sandy area, cleared and prepared for the harvest and maintenance of clams. At high tide, when the rocks of the clam garden are submerged, seagrass meadows sway in the current just beyond the clam beds. Nearby, you might find rows of other cultivated plants, including nettles, crabapples, and northwest bush cranberries. But the real marks of a well-tended clam garden, Recalma-Clutesi told us, are the sound and the smell. It smells "fresh," she says. Indigenous elders constantly turned over the mud with yew sticks, aerating the sediment and preventing the anaerobic decay smell that often characterizes tidal mudflats. And then the sound. Thousands of years of discarded clam shells crunch as you walk toward the mud. The clams squirt water from their siphons as people walk overhead, "a little bit of a symphony," Recalma-Clutesi says. "Whenever Kwaxsistalla was on his own clam garden, they'd sing to him."

"It feels good to be there," she told us. "I stand there and always am in awe of the old people who knew to extend the shoreline. They extended the shoreline on this rugged coast."

Marine ecologist Anne Salomon, a colleague of Lepofsky's at Simon Fraser University, says that the first time she canoed out to a clam garden on Quadra Island, the sound of the tide rushing through the siphon holes in the crown of the rocks suggested her future experiments to her. "You could see the activity," she says. "You can see how quickly a clam garden gets wet with the tide. It's flooded from underneath first. You can hear it, and you can see it."

Salomon arrived in British Columbia with a PhD in ecology from the University of Washington. "It became clear that the whole notion was that clam gardens made more clams," she told us. "But did they? No one had asked if there's evidence for that. That was Anne the quantitative ecologist saying, 'Prove it to me. Prove it to me how much.' I've moved rocks in the Gulf Islands; it's not easy. People would have had motivation to do it. So yeah it's a good hypothesis, but prove it to me how it happens."

Salomon and her graduate student Amy Groesbeck went looking for answers. At Quadra Island, they found perfect experimental sites. Most beaches had rock walls to create clam gardens, but some did not, allowing direct, side-by-side comparison. Salomon and Groesbeck created transects similar to what Jim Barry, Sarah Gilman, and their colleagues had done in Monterey Bay, running from the upper edge of the clam habitat to either the rock wall or, on the unwalled beaches, the average depth where the walls appeared. Paddling from site to site in canoes because Salomon's just-established lab couldn't yet afford a motor boat, they dug, identified, and measured clams at fifteen locations along each transect. In a 2014 paper, they reported that the walled gardens yielded four times more butter clams and twice as many littleneck clams as unwalled beaches nearby.[20]

Salomon and colleagues also wondered if they could show *how* the clam gardens produced more clams. As part of the same experiment, she and Groesbeck transplanted littleneck clams to different places along the transect, then pulled them up 160 days later to see how they'd grown. The clams grew significantly faster behind the rock walls and especially at the upper and lower ends of the tidal transect, the areas most affected by the wall. The gardens also provided habitat for a host of other species, including octopus and chitons.[21]

Salomon and coauthors note in their paper that "documenting these traditional practices and their ecological and societal benefits will help First Nations during a pivotal time, as First Nations continue to assert their rights to access traditional lands and resources and secure sustainable food production into the future." Much of traditional Western conservation rests on a philosophy that people are bad for nature, so the way to protect nature is to remove people: bans and regulations allow plants and animals to recover. A different philosophy underlies many Indigenous conservation approaches.[22] From controlled wildfires in California to Hawaiian fishponds, to the clam gardens of the Pacific Northwest Coast, there are ways to avoid overexploiting resources by becoming more involved and connected to the environment. People can manage habitat and harvests, and boost productivity. In doing so, they develop a closer relationship with the world around them and thus the flexibility to adapt to change.

Salomon has worked with Barb Wilson to advocate for the Haida's right to manage their kelp forest ecosystems by hunting otters, and in her research, she has advocated more generally for returning Indigenous people to their land and traditional practices as a conservation practice.[23] "Sustaining global food production presents one of the greatest environmental and humanitarian challenges of the 21st century," Groesbeck and Solomon wrote in their 2014 paper. "Fortunately, evidence from the past often offers solutions to contemporary quandaries."

The ancestors who chose Adam Dick to preserve their collected knowledge wanted him to be a conduit to the future. Deur, who worked with him for decades and who became an adopted member of Kwaxsistalla's clan within the Kwakwaka'wakw Nation, told us how Adam's teachers had perceived and chosen Adam as a "time capsule," a way of sending necessary knowledge forward into the future. Like Isabel Meadows, his ancestors might have been unsure where their knowledge would end up, but they knew it was important enough not to let go.

Most First Nations people never ceded their coastal territory, yet federal environmental regulations often criminalize traditional environmental modification or traditional harvest. In spite of this, coastal Indigenous people around the world continue to revitalize traditional practices that worked for millennia to increase the productivity of their territory and work in partnership with scientists and resource managers. Lepofsky says that it is not changing nearly fast enough, but the direction of the change is apparent.

Adam Dick died in 2018. Recalma-Clutesi says that, on trips to the clam gardens in his later years, he'd sometimes pause, think about the changing world, and say, "I'm the one. My grandfather used to be the one, now I'm the one." But the point of investing all that knowledge in him was that the world needed that knowledge, and Adam would someday be in a position to share it. Through his partnership with Recalma-Clutesi, his work with researchers like Deur and ethnobotanist Nancy Turner, and the many recordings and publications they produced together, his knowledge is available to people to make the world better now. Recalma-Clutesi told us that she is still in mourning, but one of her projects is to create short videos of Adam, speaking in the Kwakwaka'wakw language as he describes root gardens, so that children in Kingcome and other Kwakwaka'wakw communities can hear their ancestral language and see their traditional practices. "Adam often said, 'This is like going through a great flood," Recalma-Clutesi said. "'But we will come out of this. We will survive. But we will be different, as will the world. But it's not to us to sit and lament what was, but to maintain the integrity of the values of why and how we did things.'"

Shellfish have supported millions of people around the globe for the entire history of our species. Our knowledge has brought us to a point where we understand how to grow them and thrive from them. But the changes coming to the oceans in the next decades will put that knowledge to an extreme test. "We're not a bunch of eccentric antiquarians who worry about what happened 300 years ago," Deur says. "You have an obligation to not be the end point of that knowledge, to be a conduit of that knowledge into a future time. Because this knowledge is perceived as having great capacity for healing, in our communities and in the natural world."

After centuries of genocide and colonialism, it is not possible for Western societies to simply find the Indigenous science and knowledge that's survived, remove it from its community and context, and apply it to solve pressing global problems. Healing means undoing some of the harm done in the name of science in the past. It also means integrating every form of knowledge that humans have acquired, including the considerable body of scientific knowledge of the oceans, to start to heal the ocean itself and to protect it so that ideas like clam gardens can flourish. "On the Northwest Coast of North America, as in coastal communities worldwide, the human-clam relationship is age-old and continues today," a group of scientists including Lepofsky and Salomon wrote in 2019. "Tracing that history and situating these relationships in the context of modern management decisions take bringing together data from multiple sources and using diverse types of analyses. They also require recognizing the sometimes-active role of humans in modifying coastal ecosystems of the past as well as the present and that not all long-term human-ecological interactions have negative ecological consequences on biological diversity."[24]

Scholars like Melissa Poe argue that science should reorient to better understand culturally significant resources and listen to Indigenous knowledge-holders. Long-enduring relationships to the ocean, when scientists respect them, show where science can be applied to benefit people in ways outside traditional scientific practice. "I listen to the stories of people who've been watching," Poe told us.

In 2015, an ocean conservation group called the Ocean Modeling Forum asked Poe to join a working group dedicated to the Pacific herring. Part of the premise of the modeling forum, an idea shared more widely in the conservation world, is that modern conservation demands more than the classic Western model of protecting wildlife by creating regulations to protect a single species at a time. Instead, Poe says, the herring group became a way to "organize a diverse set of values and ways of experiencing a system around that single species."

Herring have significant cultural and ecological connections to the entire West Coast. It is simply impossible to separate them from the plants, animals, and people they influence. They swim in gigantic schools that often number in the millions. Whales, dolphins, seals, sea lions, birds, and bigger fish feed themselves and their families on herring. After spending most of the year off the coast, herring flood into shallow bays and estuaries to spawn in what researchers have labeled a "silver tide." The herring spawn, and each female can then lay tens of thousands of pea-size eggs in long amber strands resembling grapes. The eggs cling to the surface of seagrass blades and coat the shoreline like a bathtub ring. Huge numbers of birds and marine mammals follow, churning and frothing the surface even close to shore in heavily populated areas of Puget Sound and the San Francisco Bay. People, too, gather to catch both herring roe and the fish themselves. The herring have supported traditional and commercial fisheries for centuries. Herring have been in decline for decades, however, for reasons both local and global. So Poe's group wanted to know how to take on the topic of protecting the herring in a way that incorporated the tiny fish's impact on the wider world.

They started by trying to "just write out all the ways in which herring matter in the world," Poe says. The first meeting brought together mathematical modelers, empirical biologists, Canadian and U.S. government conservation scientists who set the fishing quotas, anthropologists like herself who work on oral histories or with traditional harvesters, and traditional practitioners with expert knowledge. Everyone told their herring stories, Poe says, until new lines of inquiry started to emerge. "Traditional knowledge tells us a thing, now we want to look at the data on herring population, and how those two are in conversation with each other," Poe told us. "If we begin to predict things from the oral history, what would that look like if we model it?" Such cross-discipline, cross-cultural collaboration is a model for how conservation can work. The more urgent the conservation challenges become, the more important it is to listen to everyone.

Skye Augustine, who has worked for more than a decade restoring clam gardens in the territory of her mother's Salish Nation people on the east side of Vancouver Island, told us that, even though more scientists now understand the value of Indigenous scientific knowledge, it's still a tough choice for Indigenous youth to pursue formal academic training. It involves merging Western intellectual traditions with cultural and spiritual knowledge, she said. It sometimes means you will later have to unlearn things. But it is also a way to bring every possible way of knowing into one place and to understand the specific language of the people who have had power over coastal policy for the last two centuries. After working unofficially with Anne Salomon for several years, Augustine joined Salomon's lab to pursue a PhD in coastal marine ecology. She works closely with Salomon, Lepofsky, and Kii'iljuus Barb Wilson as part of the Clam Garden Network, a 300-member academic government–First Nations collaboration to restore clam gardens throughout the Pacific Northwest. "What we see today is a huge disconnection between people and our oceans," Augustine said. "And we don't really have time for that anymore."

Time and tide wait for no one, as the saying goes, and climate change, overfishing, and pollution have changed the way ocean people worldwide see their work and lives. Graduate students in the marine sciences worldwide, unlike many of their faculty advisers, have never lived in a world in which climate change wasn't a critical problem. These students talk of urgency and bringing every idea and every way of knowing to solve it. We are in "a moment where we need all science on the table, right now," said Priya Shukla, a graduate student in ecology at the University of California, Davis. "I feel the pressure, the amount of transformative change that needs to happen this decade."

Shukla became interested in marine science as an undergraduate at Davis and pursued a master's degree at San Diego State University, studying the way changing ocean temperatures and pH affects kelp reproduction. She then returned to Davis as a lab technician, but she had already started to run into the often-insular culture of university research. Questions and answers seemed timed to grant cycles and scientists often investigated theoretical questions that didn't benefit people outside the university. In a world that needed urgent change, Shukla says that she grew increasingly ambivalent about academia's ability to meet the crisis. Sustainable aquaculture seemed to offer a path through. This kind of academic and industry tie was more accepted than in many other fields, in part because oyster aquaculture itself was popular on the West Coast. The idea of aquaculture appealed to Shukla, who says it helped her merge her identity as an ocean scientist and as the child of immigrants who came from farming villages in India. There was opportunity because Shukla's academic collaborators had forged a unique partnership with the oyster industry and the Hog Island Oyster Company, based forty miles north of San Francisco in Tomales Bay. Most of all, there was a clear need. Aquaculturists, including clam gardeners, work constantly to improve growing conditions for shellfish. There is, as Douglas Deur suggests, tremendous potential for healing in a closer relationship with the coast. But, like kelp and coral, shellfish are in deep trouble in the changing ocean.

"We're faced with some rather large existential challenges," says Terry Sawyer, the cofounder of Hog Island Oyster Company. Sawyer was one of a handful of aquaculturists who attended a scientific conference in Monterey, California, in 2012, on the changing ocean.[25] At a packed plenary session, then National Oceanographic and Atmospheric Administration (NOAA) administrator Jane Lubchenco challenged the assembled scientists to break down the walls between their discoveries and the communities around them. She asked scientists who among them would be willing to open their labs and field sites to members of the media, the public, and lawmakers to communicate the impacts of carbon dioxide on the ocean. Sawyer says that he watched every hand in the auditorium go up.

Hog Island Oyster Company has worked closely with scientists ever since. Shukla is part of the second generation of scientists to work with Sawyer and remembers that his first question to her was, "Can you do what you say?" Scientists can help industry, they agree, but it takes the willingness Lubchenco called for to be part of a different culture. "Everyone's competing for the same money, egos, publications," Sawyer said. "And here we are, on this worldwide problem, who's willing to raise their hand and say, 'I'm going to share my information, I'm going to work together with other researchers, this is a common goal and we don't have time for that kind of thing.'"

Between 2006 and 2008, oyster farmers at Whiskey Creek Hatchery on the Oregon coast noticed a significant mortality event inside their hatchery. They had seen unusual numbers of oysters dying of disease on and off over the last few years, but this time close to 75 percent of their oysters died even after they'd filtered out pathogens.[26] Oyster farmers typically buy oysters as "seed"— six-millimeter larvae—from hatcheries like Whiskey Creek. Although oyster farms are distributed up and down the North American West Coast, the hatcheries they all rely on are much rarer, and Whiskey Creek was one of two major suppliers.

"We had to say, we'll take whatever you can produce, whenever," Sawyer remembers. "When you have these mortality events where you can lose greater than 50 percent, that's not a way to do business. It wasn't just us. It affected the entire West Coast industry." The owners and staff at Whiskey Creek collaborated with scientists at Oregon State University to test samples of the seawater in their lab. At first, there was so much dissolved carbon dioxide in the water that they thought they'd made a mistake with the samples. But there was no mistake. They had documented an event that may have seemed unusual or poorly understood at the time, but it reflected a picture of what the future might hold.[27]

Recall that, as we put excess carbon dioxide into the atmosphere, about 30 percent is soaked up by the ocean.[28] The absorption of that carbon dioxide fundamentally changes the chemistry of seawater, making it more acidic. Animals that make hard parts, like shells or corals, construct themselves out of calcium and carbonate ions, the latter of which becomes more scarce in high carbon dioxide environments. Joanie Kleypas had first estimated what that process would mean for the reef-building corals, and the result had alarmed her so much she'd spent years hoping to be proved wrong.[29] Now, as more scientists started working on climate change in the early 2000s, a major branch of the scientific community started to think about what the math of ocean acidification would mean for other marine animals. By the time Whiskey Creek saw its oyster larvae dying in acidic water, scientists had, in parallel, begun to raise the alarm around the world and partner with sustainable fishing and aquaculture groups to look for solutions.

Oysters are often referred to as the "canaries in the coal mine" for ocean acidification, although if you are an oyster grower, that metaphor may not sit well. The observations at Whiskey Creek Hatchery, and the scientific experiments that have followed since, sent a clear message to scientists and policymakers that ocean acidification is going to wreak havoc on coastal economies, communities, and food sources with deep historical and cultural roots. Among the first observations about ocean acidification on groups like oysters, mussels, and clams was that, under elevated carbon dioxide conditions predicted for the future, these shellfish exhibited smaller or thinner shells, making them more susceptible to predators and the harsh marine environment.[30] With more research, scientists have been able to document far-reaching effects beyond shells: ocean acidification may fundamentally alter the physiology and reproduction of many organisms, predator-prey behavior, and competition between organisms.[31] These impacts have been shown in a wide variety of taxa and environments around the world, including fish, corals, kelps, and snails.[32]

Being a farmer or a fisher in any era requires a certain connection to the Earth, the weather and seasons, and the cycles of life. Commercial shellfish farmers build their business and their lives around tides, movements of water, and seasons; atmospheric rivers that bring deluges of freshwater are as consequential as an unusually warm spring or a weak upwelling, and each demands close attention. Ocean acidification is a universal background condition that will make it harder for shellfish to survive anywhere on Earth.

Hog Island Oyster Company's Terry Sawyer grew up near Cape Canaveral on the east coast of Florida and describes his childhood as an amphibious one. He explored, swam, sailed, fished, surfed, and ate from the sprawling lagoons of the Indian River. He also started working with researchers early. As a high school student, he found a job at the nearby Gulfstream facility, where engineers tinkered with a deep-sea submersible, the *Johnson Sea Link*. In the late 1970s, Sawyer finished an undergraduate degree in marine biology at the University of California, Santa Cruz, and moved to the opposite shore of the Monterey Bay to join the husbandry staff of the start-up Monterey Bay Aquarium. For six years, he learned a bit of everything, he said, about rocky shore communities, cephalopod care and feeding, how to build and maintain a trout stream habitat, sandy shorelines, sea otter rehab. He collected shark specimens, fixed tanks, worked with veterinarians, and captured and released sea otters and jellyfish. Then in 1988, two friends asked him to put his experience to work in building an oyster aquaculture farm on five acres in Tomales Bay, California, and Sawyer accepted.

There are perhaps more than 100 species of oyster in the world, and scientists believe they have lived on Earth for at least 250 million years. The exact number of species isn't known because oysters are notoriously difficult to identify and classify, even for trained scientists. Most of what people see of an oyster is its shell, and oyster shells vary enormously, even within a single species. They also seem to be spinning off new species more rapidly than many other creatures, leading to a tangled taxonomic tree.[33] Most North American aquaculture, however, focuses on a handful of common oyster species, especially the Pacific oyster. When Sawyer started as an oyster farmer in the 1980s, he told us, the business challenge was in some ways quite basic: convince West

Coast consumers to eat a smaller Pacific oyster than what they were accustomed to purchasing from the East Coast.

By and large, they and their aquaculturist colleagues have succeeded. Oyster farming is a growing industry in the United States, part of a larger shellfish farming community that contributes about $300 million annually to the U.S. economy. Hog Island Oyster Company itself has expanded to farm 160 acres and operates the first permitted oyster hatchery in California. With a market established, Sawyer said, Hog Island Oyster's concerns started to mirror many other businesses: dealing with permits and leases, the high cost of living on the California coast. Now, as Sawyer deals with more typical business concerns, he is also thinking about the changing ocean and how this industry will adapt.

Sawyer and his crew typically spend their days in waders, mucking around in small boats in the fog of Tomales Bay to maintain their baskets of oysters growing to market size in the cold Pacific waters. In the last decade, he has become more comfortable in scientific and policy settings. Sawyer and his colleagues walk the halls of the state capitol to talk with lawmakers about the effects of ocean acidification on local businesses and communities. Many aquaculture growers have become immersed in the combined work of understanding the science of ocean change and communicating these threats to their customers, elected officials, and neighbors.

Sawyer has joined with scientists and other aquaculturists to set up monitoring stations along the coast that would inform both the scientific community and the shellfish farmers about current ocean conditions. These monitoring instruments are like the weather stations of the sea—providing rapid, frequent observations of ocean temperature, salinity, and carbon dioxide. Access to scientific information provided opportunities for oyster farms to make different decisions—when to turn the water pumps on, or put the young oysters out—that helped their business adapt to the threat of rising carbon. Farmers like Sawyer rapidly became experts in ocean chemistry, and scientists at many coastal universities became engaged in conversations with shellfish farmers about how the discoveries on ocean acidification could help them plan for their business.

Still, the threat persists as long as carbon dioxide concentrations continue to increase in the atmosphere, and the challenges become more complex.

A kelp forest can be directly susceptible to warming and also vulnerable to a new grazer that thrives in the absence of a predator. A shellfish bed or a shellfish farm, likewise, has layered vulnerabilities. Priya Shukla started running ocean acidification monitoring stations, but more recently, her work with Hog Island Oyster Company has focused on a related problem: oyster disease. As the COVID-19 pandemic made clear, viruses can exploit weaknesses where they find them. A changing ocean, as it turns out, may weaken oyster disease resistance, making oysters more vulnerable to viruses and pathogens. Shukla's experiments are set up as oyster "boot camps" that test to see whether young oysters exposed to warm conditions in the lab can then be more resilient to heat stress and disease in the estuaries where they are grown. Results show a promising trend: young oysters exposed to temperature stress at an early age may actually be stronger in the wild when they are exposed to diseases and fluctuating temperatures.[34] Commercial oyster farms might be able to use a similar approach to "climate-proof" their harvests. At some point, however, oysters and all shellfish are like corals: they cannot run from warm weather. In the last few years, warming air temperatures fueled by climate change coincided with summer low tides, roasting shellfish in place across the Pacific Northwest and Northern California. The ecological effects of these die-offs will linger for years.[35]

Can the changes happening in science and conservation happen quickly enough to address today's problems? The next time we step out onto the mudflats, whether it is in British Columbia or Tomales Bay, will the crunch of the shells beneath our feet remind us of millennia of knowledge that humans hold about the way to live in partnership with the resources of the sea? Clam gardens and oyster beds—and changing tide pools, coral reefs, and kelp forests—can be hard to notice at first for people who aren't at the ocean every day. In some sense, Dana Lepofsky says, one of the biggest challenges of all is getting people to notice. "If you don't see it, you don't see it," she told us. "I'm doing tours all the time, and people are just stunned. I hear this over and over, 'You've changed my view of my home, of my landscape. I had no idea.' Things are really obvious to us as scientists, or to Indigenous peoples, [but they] are not to the public. That's part of the problem."

A new wave of hope and innovation is hitting the aquaculture industry today. New farms are opening that aim to harvest both seaweed and shellfish, in some cases side by side. The seaweed, which removes carbon dioxide from the surrounding seawater, may make the habitat more conducive for the growth of the nearby shellfish. Women aquaculturists and Indigenous groups are focused on developing farms at a time when demand for shellfish is on the rise, and farmed shellfish and fish is predicted to play an important role in providing protein from the sea. "There are many on the ground bright spots with Indigenous practices happening in [British Columbia]," Anne Salomon told us. "Clam gardens are a beautiful nexus of this. There is a huge legacy of intergenerational power loss and knowledge loss. Clam gardens are a way to recapture this."

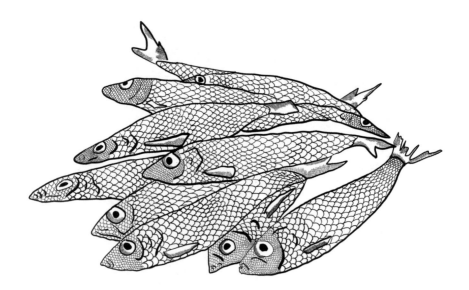

CHAPTER 5

THE ABUNDANT OCEAN

It reminds me that if a fishing company can do its inner and outer work on a national scale, I can most certainly remember the connection between my inner transformations and my actions in the world. I can, you can become more than a market. I can, you can remember that the world is round. What we touch, how we steer, the roundness is the measure of our purposeful living. The impact is always beyond one species.

—Alexis Pauline Gumbs,
Undrowned: Black Feminist Lessons from Marine Mammals

Kawika Winter did not intend to get a PhD in science, did not intend to work in fisheries management science, and certainly did not intend to become a researcher running a program funded by the U.S. government. As he neared high school graduation on the island of Oʻahu and peers talked about which elite mainland colleges they would be attending in the fall, Winter had no interest in leaving Hawaiʻi for "a piece of paper that said I was an educated person." He went instead to the University of Hawaiʻi to continue learning about his home and community. As an undergraduate student, when teachers told him he should consider pursuing science and graduate degrees, Winter would reply, "I don't want to be stuck in this concrete building with you guys, I want to be on the *ʻāina* ['the land']." Then, as an activist, he watched from outside the building as the PhDs and the lawyers went inside to make decisions. So he joined a PhD program as a way to gain influence in those closed-door sessions, only to find a job almost immediately restoring an *ahupuaʻa*, a mountain-to-sea landscape slice arranged in the traditional Hawaiian way, at the Limahuli Garden and Preserve on the island of Kauaʻi. That project to return Native Hawaiian management to the land was exactly what he had long wanted to do, so planned to spend the rest of his life on it and ultimately die on his restored land. Instead, after more than a decade on Kauaʻi, his daughter won a scholarship to a high school on the island of Oʻahu. Winter, who had ended up finishing that PhD while working in the botanical garden, saw a suitable job posting for a research director of the U.S. government's National Estuarine Research Reserve (NERR), in the ahupuaʻa of Heʻeia on the east shore of Oʻahu.

In one of the remnant patches of tropical forest in the rugged hills overlooking Kāneʻohe Bay, Heʻeia Stream trickles down into the nearby town. Its narrow green ribbon flows through the new ecosystem of developed Hawaiʻi: past suburban houses and under the Kahekili Highway, past more houses and the parking lot of the Church of Jesus Christ of Latter Day Saints, before emptying into a wide estuary on the coral-studded Kāneʻohe Bay. There, where the fresh water and saltwater mix, Hawaiians have raised fish for the last 800 years. The eighty-eight-acre Heʻeia Fishpond is enclosed by a 1.3-mile long, arcing coral-and-basalt wall. Sluice gates on the ocean side allow its managers to control both seawater and fish coming and going. Three more gates on the land side divert the stream into the right parts of the pond and allow a fine-tuning of the water salinity, which in turn controls the rapid growth of the algae that fish like to eat.[1] Agricultural runoff further enhances the nutrients in the pond. Young wild fish enter from the ocean side, grow old and fat on abundant food and lack of predators, and then can be harvested or released to spawn in deeper water. Like a clam garden, a fishpond is an Indigenous way of increasing abundance to benefit people and nature alike, and several clam garden practitioners told us they had traveled to Hawaiʻi to learn from the people who were restoring fishponds. Heʻeia Stream was part of "a system unique in the world: hybrid, cultivated-wild aquaculture using ponds to trap, raise, and harvest ocean fish," journalist Erica Gies wrote in bioGraphic in 2019.[2] Until sustained European contact in the late 1700s, there were at least 488 such fishponds in the Hawaiian Islands and more than thirty in Kāneʻohe Bay alone, each of them connecting to upstream agriculture in a mutually beneficial relationship for crops, fish, and people.

American colonists arriving in the eighteenth and nineteenth centuries didn't value the benefits of the fishponds, which suffered a similar fate as the clam gardens.

As agricultural plantations spread across the islands, trade thrived on imported food from the continent, and as new settlers made a determined effort to supplant traditional Hawaiian systems of land tenure, the fishponds fell apart. The Heʻeia estuary suffered the fate of so many coastal zones worldwide: as the islands became more developed, the estuary grew more polluted and invaded. Sediment had run downstream, changing the depth of the fishpond and damaging its walls. The force of floodwaters broke through rock walls and damaged the sluice gates. Although the outline of the pond remained, it no longer served its original purpose.[3]

By 2001, after decades of battles led by community elders to protect the area from development, a group of emerging Native Hawaiian leaders in the community intervened. A local nonprofit called Paepae o Heʻeia formed and started working to restore the fishpond and the community around it.[4] Despite a long legacy of distrust of the state and federal governments, the group of elders and young community leaders also saw an opportunity to try a new form of partnership to help bring funding and visibility to the estuary while maintaining local control. They asked the state of Hawaiʻi to nominate Heʻeia for inclusion in the U.S. National Estuarine Research Reserve (NERR) system, which encompasses thirty coastal sites across the United States and its territories. In January 2017, President Barack Obama signed an order to create the 1,300-acre Heʻeia NERR.[5] The NERR would be a collaboration between Indigenous community organizations, and state and federal governments, with a governance board that included Native Hawaiian leaders. Its mission would go beyond research and restoration to revitalizing the community living around it. A few months after its creation, the reserve searched for its first director and Kawika Winter, convinced by its grassroots origins, took the job.

If Winter has found himself in an unlikely position relative to what he once intended, his role as director of the Heʻeia reserve has given him the ability to leverage his deep knowledge of one of the planet's most biodiverse island chains for what he believes will be the good of all humanity. In an article in *American Scientist* in 2019, Winter and University of Hawaiʻi senior scientist Sam ʻOhu Gon called their vision "A Hawaiian Renaissance That Could Save the World."[6]

"The institutional education I was receiving basically conveyed the idea, 'Humans are bad, humans are the problem, humans are destroying everything,'" Winter told us. "I always knew it wasn't right. What about fishponds? Fishponds created super abundance in the ocean. How is that destroying things when fishponds are making more abundance than the ocean could make on its own?"

To make more than the ocean or land could make on its own has always been the promise of farming. And people have relied on fish farming for almost as long as they have relied on domesticated land animals. Archaeologists have suggested that fish were first domesticated in China more than 8,000 years ago.[7] In the Budj Bim Cultural Landscape in southeastern Australia, radiocarbon dating shows that the Gunditjmara people built stone fishponds more than 7,000 years ago.[8] Wall art in 3,500-year-old Egyptian tombs shows managed Nile tilapia. The oldest written work on aquaculture, *Yang Yu Jing*, appeared 2,500 years ago.[9] Efforts to farm fish also rapidly acquired a sophisticated understanding of ecology. In the 2010 book *Four Fish*, author Paul Greenberg writes that the first aquaculture practiced by ancient Chinese farmers incorporated a mutually beneficial food web.[10] Carp gathered beneath mulberry trees where silkworms spun their webs, carp feces boosted the growth of rice and other grains, and the grains fed ducks and poultry. Modern aquaculturists have rebranded such ancient ideas as "integrated multitrophic aquaculture" to raise fish, shellfish, and other marine species in an environmentally balanced way.[11]

Like clam gardens, fishponds might sound to a Western ear like a provincial solution to the overwhelming environmental crises facing the planet. But here's the statistic that Winter and other scientists share: modern estimates suggest that 300 years ago, prior to the arrival of European diseases, the population of the Hawaiian Islands was around 700,000 people.[12] These Hawaiians lived entirely on what they harvested from the islands; University of Hawaiʻi researcher Natalie Kurashima estimated that, at that time, Hawaiian terrestrial agriculture alone produced enough food for as many as 1.2 million people.[13] Today, the population of Hawaiʻi is around 1.4 million—but between 85 and 90 percent of its food is imported. Nearly 50 percent of its seafood comes from outside the islands.[14] Rare plants and animals have gone extinct, once productive forests and agricultural lands have been destroyed, and although there's a burgeoning fishpond revival sweeping through places like Heʻeia, only fourteen of the fishponds that once generated 2 million pounds of protein per year remain in production.[15]

Winter says that this leads to an obvious side-by-side comparison. His ancestors, working closely with the land and ocean, managed a system that fed nearly 1 million people without overdrawing the ecological balance of the islands. Modern Hawaiʻi can only feed a comparable number of people by bringing the food in from elsewhere, and the fish, the coral, and the land are worse off.

Western conservationists see a crisis of scarcity on the islands today and call for increased regulation and protections for what remains. But Winter says that it is not purely more protection that is needed. He says that the world needs a cultural shift toward seeing nature the way his ancestors did, as something integrated into human culture, and the world needs a political shift that puts people with that worldview in power.

"We can tap into that memory, we know what used to be here, we know how to do it, if you just let us do it," Winter says. "Our ancestors have gone through island ecosystems that are extremely limited, and came out of that with an abundance mindset. Continental people come to Hawai'i and they're stuck with a scarcity mindset. How did our ancestors come up with an abundance mindset?"

If many people now see the ocean as a place of scarce and fragile resources, it's perhaps a development born from the collapse of an earlier view that fish were infinite. In an 1883 address to the International Fisheries Exhibition in London, the influential English biologist Thomas Huxley told the assembled dignitaries, "I believe that it may be affirmed with confidence that, in relation to our present modes of fishing, a number of the most important sea fisheries, such as the cod fishery, the herring fishery, and the mackerel fishery, are inexhaustible. And I base this conviction on two grounds, first, that the multitude of these fishes is so inconceivably great that the number we catch is relatively insignificant; and, secondly, that the magnitude of the destructive agencies at work upon them is so prodigious, that the destruction affected by the fisherman cannot sensibly increase the death-rate."[16]

Huxley admitted in his talk that an English river here or there might see its salmon extirpated through the fault of pollution. An oyster bed might be dredged clear, and the oysters driven away for good. In such cases, however, the solution was clear: "Man is the chief enemy, and we can deal with him by force of law." If Huxley's prediction about the number of fish in the ocean had been staggeringly off the mark, this vision for how Western societies would take conservation action was prescient. In 1883, he staked out some of the contours of the nongovernmental organization (NGO)–versus–industry binary that still persists today: "man" is the problem. If fish are threatened, the obvious solution is to outlaw fishing.

In 2010, the signers of the UN Convention on Biological Diversity agreed to a goal of protecting 10 percent of their territorial water by 2020. Although small-scale marine refuges had existed for centuries or longer, and many of the world's most iconic marine parks had been established by the 1990s, the first decades of the twenty-first century saw the creation of massive new marine reserves aimed at increasing the percentage of protected ocean. Even the United States, the only country in the world that hasn't ratified the Convention on Biological Diversity treaty, grew its area of protected ocean from 1 percent at the start of the century to 26 percent by 2022.[17] Its new marine protected areas included the Papahānaumokuākea National Monument off the northwestern Hawaiian Islands, created in 2006 by President George W. Bush and expanded in 2016 by President Barack Obama to cover 582,578 square miles, an area nearly as large as the Gulf of Mexico. As an indication of its sheer size, Papahānaumokuākea is nearly twenty times larger than the largest terrestrial protected area in the United States. In the same year as the Papahānaumokuākea expansion, the United Kingdom roped off a 320,000-square-mile protected area around the Pitcairn Islands in the South Pacific, and governments from around the world, including the United States and European Union, agreed to boundaries for the 598,000-square-mile Ross Sea Region Marine Protected Area in Antarctica. In 2018, the Australian government created the 382,000 square mile Coral Sea Marine Park to protect a vast area beyond the Great Barrier Reef.

As the percentage of the global ocean under some form of protection grew, scientists, conservationists, and some Indigenous groups began to push back against what had started to look, to them, more like a competition to outline large areas on a map. Scientists became concerned that the size of the reserves didn't necessarily equate to protection and that many reserve boundaries were drawn to avoid areas most affected by fishing or mining. They wondered about enforcement in vast areas of ocean thousands of miles from continental population centers. And they worried that the focus on a few big areas might actually undermine conservation.[18]

Indigenous Pacific Islanders, meanwhile, saw massive reserves announced for the Pacific by Western governments seated far from the protected areas. The boundaries had often been drawn without input from the people they actually affected. As the Mariana Islands native and conservation scientist Angelo Villagomez wrote of his homeland in a 2008 blog post, "Sometime in the last 20 years, locals in the CNMI [Commonwealth of the Northern Mariana Islands] came to understand conservation as a bunch of haoles telling Indigenous people not to fish, not to feed their families, and not to practice their culture."[19] Villagomez says that he thinks this is a misconception. He has become one of the leading advocates for marine protected areas (MPAs) around the world and for a goal of protecting 30 percent of the world's oceans by 2030. Yet he points out often that it is much harder to make the case for MPAs when planning efforts don't include everyone. The overwhelming majority of federally protected waters in the United States lie in the Pacific Islands, Villagomez told us, but he has searched for other Pacific Islanders on the staff of the many well-funded ocean conservation organizations in Washington, DC, and found "just two of us." "Conservation has not included Indigenous people in our movement," he says. "It's a huge missed opportunity. Especially when you look at the data on how well Indigenous communities do in protecting nature."

When people first arrived in many of the Pacific Islands, rapid environmental change followed. A 2013 estimate using fossil evidence from forty-one of the most recently settled islands argues that more than 1,000 species of birds were driven extinct.[20] Settlers burned forests, likely for hunting and to clear space for horticulture, and historian Madi Williams estimates that nearly half of Aotearoa New Zealand was deforested by the time European settlers arrived. The origin of the Indigenous Pacific Islander conservation ethic starts with that initial overexploitation, Williams concludes: "It was soon understood that the resources were finite and efforts were made to conserve them."[21]

Pacific Islanders responded with management systems that benefited people and protected biodiversity. "Almost every basic fisheries conservation measure devised in the West was in use in the tropical Pacific centuries ago," tropical marine ecologist Robert Johannes wrote in 1978 after spending a year in Palau.[22] Each cultural tradition has its own concept of protection: *mo* in the Marshall Islands, *bul* in Palau, *tapu* in New Zealand.[23] "You can pick an island and ask someone what their concept of a protected area is, and they'll be able to give you the word in their own language," Villagomez says. Kawika Winter, Villagomez, and many other Pacific Island scholars use the metaphor of the voyaging canoe for their stewardship of the islands. Winter says that there is a Hawaiian proverb:

"A canoe is an island. An island is a canoe." The canoe had to carry everything its people needed for months at a time. So, too, did the islands.

Villagomez told us that the COVID-19 pandemic and U.S. national reckoning with racism in 2020 moved people in conservation to change the way they define protection. Percentage targets are useful, he agrees, and protecting 30 percent of the ocean is a worthwhile goal. But how protected areas are added to the total matters. "Conservation is people," he said. "I think for the next 10 years, large conservation organizations will spend a lot more time thinking about how those actions affect the lives of people who have to live with those decisions."

Carlotta Leon Guerrero, the founder and director of the Guam-based Ayuda Foundation and a former Republican senator in the Guam legislature, wrote in 2018 that Pacific Island communities were already leading the way in developing locally organized and managed MPAs.[24] For example, the Palau National Marine Sanctuary restricts fishing from within 80 percent of Palau's national waters. Other Pacific communities are banning industrial scale fishing while maintaining artisanal fishing rights. "Pacific leaders have acted with the same boldness that inspired our ancestors to cross the ocean," Leon Guerrero wrote. "We have taken their ancestral knowledge and expanded upon it, designating vast ocean sanctuaries, which support healthy marine ecosystems and abundant fish populations, while ensuring the well-being of coastal communities."

A fishpond is an ancestral technology that benefits fish in a local area. But the lessons extend beyond the shoreline, out into the fishing grounds where most of the world's fish live and fishing happens. Recent studies have shown that industrial-scale fishing affects 55 percent of the global ocean.[25] Even as massive fishing fleets deploy the latest technology to chase an increasingly scarce resource, the important questions about people and nature are the same as they are in an eighty-acre nearshore fishpond. Whose resources are these? What is the best way to conserve them for future generations? Who benefits from protection, and who is left out? A tuna, as Paul Greenberg writes in *Four Fish*, might be a "stateless fish, difficult to regulate and subject to the last great gold rush of wild food." As is true every time there is a rush for gold, however, it is worth noticing where the gold lies and where the people who rush to it come from, who profits from the gold, and who is being shoved aside.

In 1883, Huxley spoke of the way the English fishing expansion around the globe had also meant the "dissemination of the germs of civilization."

The English and their contemporary European colonial powers spread their extractive fishing technology and their philosophy that there were always more fish to find. Now as the consequences of ever-more extraction mount, and the limitations of protection alone become apparent (for fish, as they are for coral reefs), many Western conservation groups have started to look to other cultures for answers. Kawika Winter says that the Heʻeia NERR works with reserves across the country to offer advice from "the fourth decade of a Hawaiian Renaissance." From Lake Superior to coastal Alaska, the Hawaiians have teachings to share.[26]

In interviewing people for this book, we have asked Western scientists, conservationists, fishers, and Indigenous scientists and knowledge holders, in different disciplines and from different backgrounds, if they are surprised at the state of the modern ocean, at how degraded it has been allowed to become. Winter let out a sardonic chuckle and answered before we finished the question: "No. We've been saying this was going to happen for generations," he told us. "You can't take, take, take and expect that there's going to be something in the future. That's not the way it works. We've learned that lesson in island systems thousands of years ago."

The first anatomically modern *Homo sapiens* walked in Africa sometime around 300,000 years ago. But did they think or act like modern humans? Were they, as archaeologists put it, *behaviorally* modern?[27] To try to find some area of common ground in the murky world of our species' ancient past, researchers who study human origins have tried to agree on a list of evidence from the archaeological record that's consistent with the kinds of behaviors we think are unique to people. This list tilts toward the ocean: shell beads imply the capacity for art and symbolic thinking;[28] fossilized fish bones altered by human cutting imply complex thinking and the technology to capitalize on it.[29] Fishing in the ocean for wild fish, archaeologists tend to agree, is one of the hallmarks of modern human behavior.[30] It is a skill that people picked up early, got good at quickly, and took with them as they moved around the world.

By 50,000 years ago, people had settled the Australian continent. Although sea levels were lower, exposing a vast shelf and numerous islands around southeast Asia and Australia, the human expansion out of Asia and into modern-day

Indonesia, Timor, Papua New Guinea, and the Solomon Islands still meant crossing open ocean.[31] These ancestors must also have recognized the capacity of their boats to take them out into the watery grounds of the fish. In places where the modern shoreline of the islands still compares roughly to the ancient shoreline, sea caves hold fish remains dating back tens of thousands of years. At one cave, Asitau Kuru, on the eastern edge of East Timor, archaeologists have found a 42,000-year-old layer whose fish bones are 49 percent from open ocean species—tuna especially, but also jacks, sharks, and rays.[32] Many of the bones appear to come from young fish, and the authors say they think it is more likely the fish were caught with nets than with hooks and lines. But it is clear they were caught offshore. It is the oldest evidence found to date for open-ocean fishing.[33] Elsewhere in the same cave, archaeologists found the point of a snail-shell fishhook 16,000 to 23,000 years old. Although it was likely used for inshore fishing, this is among the oldest known fishhooks. Carved from the shell of a Trochus snail, it is roughly the size you would use today for larger baits like mackerel or sizable plastic worms. It doesn't appear particularly menacing, but perhaps in an ocean teeming with more fish than the modern one, the technology didn't need to be as effective.

"I'm always amazed that people were able to catch fish using the hooks we found," said Sue O'Connor, an Australian archaeologist who led the East Timor project, in an email. "They are beautifully made but they don't have the sharp jabbing end and barbs that we associate with modern metal hooks. However, as the sites have tens of thousands of fish bones and not much else it appears that the hooks must have worked quite efficiently."

A few years after O'Connor's team found the fishhook in East Timor, Japanese researchers working at Sakitari Cave on the island of Okinawa found a series of circular Trochus-shell fishhooks, which they dated as 23,000 years old. The cave itself seemed to have been occupied 30,000 to 35,000 years ago and then nearly continuously lived in for the next 25,000 years. The researchers wrote that fishing might explain how people were able to make a permanent home on a rocky, remote island with few other food resources. The circular shape of the hooks, a common shape even now for open-ocean fishing, implies sophisticated fishing. And the date itself exploded the idea of fishhook technology arising recently. Instead, the authors conclude, by 35,000 years ago "successful maritime adaptation"—the ability to live well off the ocean's resources—"was more widely distributed along the 8,000-km-long southwestern Pacific coastal region."[34]

Ancient people also started to express the meaning they found in that connection. In 2014, O'Connor and Indonesian archaeologist Mahirta started digging in a cave called Tron Bon Lei on the Indonesian island of Alor. Six feet down, in a layer roughly 12,000 years old, a human skull emerged, resting on its cheek, with the rest of the body embedded in the edge of the pit. Directly beneath the person's chin, someone had placed a fishhook and a perforated cockleshell, showing evidence of polish and still bearing traces of a red colorant. Around the skull lay four more fishhooks. It was clearly a burial site, and the fishhooks were clearly burial goods, O'Connor wrote in 2017, making this "the earliest excavated evidence for a direct link between cosmology and fishing." Tests later revealed the skeleton to be that of an adult woman.[35]

The boat, hook, and line worked by those ancient Pacific Islanders would be familiar fishing tools around the world for millennia to come. The boats got better; the fishhooks switched from shell to bone to metal; the line became finer. Yet even in 1883, as Huxley was telling English aristocrats about the inexhaustible oceans, the primary method of fishing he described was still a group of men or women going out in a boat, dragging hooks or nets behind them, and returning with fish.

In the best-selling book *Cod*, author Mark Kurlansky describes the English arriving—in their sailboats, with their fishing lines—in the North Atlantic fishing grounds of the Grand Banks in 1497.[36] (Kurlansky writes that Basque fishers had likely already been fishing the Grand Banks, but unlike the colonial English, preferred to keep the place a secret.) The European discovery of prolific cod schools off the North American coast soon led to what Kurlansky describes, again, as a gold rush. But even as New England towns grew wealthy on the European cod trade, the technology for catching the fish remained conservative, Kurlansky writes. Once they knew where the fish were, "Fishermen then pursued cod in much the same way for the next four centuries."

Technological innovations arrived, as they often do, in a batch. The combustion engine replaced sails, expanding the range of the fishing fleet. Coal- and then oil-powered metal ships were strong enough to drag newly adopted gill nets across the ocean floor and recover a multi-ton catch, and they were now large enough to deploy another twentieth-century invention: freezing. "During World

War II the three innovations—high-powered ships, dragging nets, and freezing fish—had come together in the huge factory ship," Kurlansky writes. These factory ships, fishing twenty-four hours a day, scraping the surface from the ocean floor and leaving utter destruction in their wake, led to monstrously increased commercial catch. Spotter aircraft that had hunted for submarines in the war now hunted for fish. Sonar helped the boats zero in on the exact spot to trawl. The ships got bigger, the catch got bigger. It was, Kurlansky writes, the "golden age of long-distance net trawling."

Then the crash. The Pacific sardine, the largest fishery in the United States in the 1930s and 1940s, with a peak catch exceeding 700,000 metric tons, started to decline in the 1940s. By the 1960s, the catch had shrunk by more than 96 percent, and the fishery was closed for the next two decades. (After the closure, commercial fishing started again, rapidly peaked, and then collapsed again, leading to another moratorium and fishery closure.) In the world's largest fishery, Peruvian and Chilean fishers caught more than 10 million tons of anchoveta off the South American Pacific Coast annually from 1960 to 1970. The anchoveta population suddenly and utterly collapsed in 1972; for most of the next decade, the fishers switched to sardines.[37] The average weight of swordfish caught in the mid-Atlantic declined by 64 percent from the 1960s to the 1980s as fishers caught the larger, older fish and then moved on to catching younger and younger ones.[38]

From the tide pool to the kelp forest, the dramatic changes sweeping through the world's oceans can be hard to see. We can't tell what's happening in fish spawning grounds or how much more effort it takes to get popular fish to the market. A change in the distribution of fish size isn't visible except through statistics. Small differences in ocean temperature, salinity, oxygen concentration, and acidity aren't "as visually striking as the storms and wildfires on land," two marine scientists wrote in *Science* in 2022.[39] Yet it all adds up to an ocean beyond the modern human experience. When paleobiologists look back in time at oceans that mirror what might happen in the coming decades, they interpret evidence for radically different places, supporting smaller, less palatable fish.[40]

In 1989, the Canadian government minister for fisheries issued a last gasp for Huxley's doomed concept of the inexhaustible ocean, declaring that the cod would be fine and labeling scientific advice to institute quotas "demented."[41] Three years later, the same government minister announced a moratorium on fishing for cod. In 1994, the Canadian moratorium was expanded and the United States joined it, closing portions of the heavily fished Georges Bank after surveys showed groundfish stocks collapsing.[42] "Just three years short of the 500-year anniversary of the reports of . . . men scooping up cod in baskets, it was over," Kurlansky writes. "Fishermen had caught them all."[43]

The book *Cod* came out in 1997, five years after the first Canadian closure, when the fish showed no signs of recovery. Fishers interviewed in the book optimistically expected the fish to return soon, while some of the more pessimistic scientific sources said it might take many more years. But even with strict fishing quotas, new problems arose. Warming temperatures and shifting currents, associated with climate change, have shrunk cold-water habitat and hampered the cod's ability to grow and reproduce.[44] Twenty-five years later, the fish have not yet fully recovered.

From the very beginning, the western map of ocean fisheries has been blurry. Fish were first assumed to be inexhaustible because there was so little evidence that they were not. Once it became apparent that individual fish stocks might be finite, the ocean itself remained a seemingly limitless place. In a 2002 history of U.S. fisheries policy titled *From Abundance to Scarcity*, author Michael Weber tells the typical four-part story of a kind of anglerfish called the monkfish: at first an unwanted bycatch; then introduced to the world by Julia Child as a desirable dinner, just as other species were becoming overfished; then rapidly overexploited until the fishery collapsed; and then a slight recovery as a wave of conservation measures swept through fisheries policy in the 1990s.[45] When no one really knows how many fish are out there, but some people definitely know they can make a few dollars from selling a fish, it tends to make for an asymmetrical debate over whether to fish or not. "In this atmosphere," Weber writes, "fisheries developed with no idea of how large or productive fish populations were, and generally faced restrictions only when problems became inescapable." If there has been a theme to Western interaction with the ocean, it's this asymmetry. Western scientists, Indigenous scientists, and conservationists try to delay or push back against resource extraction in the absence of understanding, while industry and governments press ahead. Now as new technology uncovers previously inaccessible areas of the ocean, as our maps get bigger and more precise, the same tension continues to play out, from nearshore fisheries to the poles and the deepest reaches of the abyss.

The oceans are capacious places. So large that, for most of human history, it would have seemed nonsensical to try to quantify how much fishing actually takes place and where. But of late, our capacity for making sense of truly massive data has increased. In 2018, a group of scientists led by the conservation NGO Global Fishing Watch had an idea that they could try to quantify our footprint on the oceans using a modern big-data approach.[46]

Under a treaty called the International Convention for the Safety of Life at Sea (SOLAS), almost every boat heading out to sea now comes equipped with what is called an automated identification system (AIS).[47] The AIS broadcasts information about the boat and its position every few seconds, helping to avoid collisions. Each of these pings is also recorded by satellites, making the location data available worldwide. For four years between 2012 and 2016, the Global Fishing Watch research team used 22 billion AIS positions to train a computer neural network to recognize the distinct signatures of fishing boats. The computers learned to tell the difference between fishing and other types of boats, and could distinguish, with 95 percent accuracy, between types of fishing boats like trawlers, longliners, and purse seiners.

Once the computers had learned to make sense of the data, the researchers asked them to track what they saw. In 2016, the computers followed 40 million hours of fishing on the ocean. To try to estimate the extent of global fisheries, the researchers divided the oceans into a grid of roughly thirty-five by thirty-five square-mile cells and counted any cell as fished if a boat spent time fishing in it. The result, they reported, was that more than 55 percent of the cells were fished. When they tried to account for areas with poor satellite coverage or places where some boats still don't use AIS systems, they wrote, a "generous assumption" would be "that 73 percent of the ocean was fished in 2016."

Of course, a lot of estimation is involved in those percentages, from the computers' use of the AIS data to the grid itself. Critics point out that counting an area more than twice the size of Los Angeles as fished because a single fishing boat visited it once in a year would likely grossly overestimate the effect of fishing on the oceans.[48] Nonetheless, it's clear that in the twenty-first century, there are few places, even in the farthest reaches of the endless ocean, that don't see, at the very least, the occasional human visitor. Since the first Southeast Asian and Pacific Islanders dropped their hooks into the deep blue tens of thousands of years ago, we have figuratively shrunk the oceans by an unimaginable amount.

Following the big fisheries crashes of the 1980s and 1990s, governments worldwide switched to more conservative fisheries policy. A decade later, some of these policies seemed to have worked. Swordfish bounced back in the Atlantic. In California, rockfish and sardines returned in numbers sufficient to support a commercial fishery once again—although only temporarily for the sardines. As the anecdotes of recovery trickled in through the 2000s, scientists started to debate just how well the world's fisheries were doing. In the early twenty-first century, we had filled in our map of the ocean to the point that scientists started to find it possible to write a single paper with information from every part of the world. Right around the same time that research teams were analyzing coral reefs, seagrass beds, and kelp forests worldwide, a group of fisheries scientists decided to summarize everything they could gather about the state of the world's fisheries. They examined every data source they could find, from every angle they could conceive. They used stock reports, catch reports, research surveys, and ecosystem models "organized hierarchically like a Russian doll," the scientists wrote in 2009.[49] The conclusion was that the ocean remained a "mosaic of stable, declining, collapsed, and rebuilding fish stocks and ecosystems." Conservation had worked in some places and had not been attempted in others. Ending on a hopeful note, the authors wrote of their vision for a world unified—despite the short-term costs—around the theme of sustainable marine fisheries.

One person inspired by this theme was Malin Pinsky. In the 2009 paper, the word "climate" is used exactly once, under the heading "open questions." Pinsky, then a graduate student at Stanford, thought he might be able to take a similar all-available-data approach to start answering the open question about climate. Like Jim Barry and colleagues in the tide pool a decade earlier, researchers had just started to consider what climate change might mean for open-ocean animals. It was becoming increasingly evident that global change would be the story of the future. "I feel like I went through school at a time when the impact of fisheries on the ocean was really becoming clear," Pinsky told us. "The ocean was being transformed by human activities, especially through fishing. The idea of change, in that sense, was sort of built into my thinking from the beginning."

Pinsky's graduate work explored marine species dispersal, ranges, and habitats and even documented differences between the hunting or fishing and management of terrestrial versus marine species.[50] As he began his career as a professor, Pinsky continued to wonder what role climate would have in the trajectory of fisheries management. He knew that scientists had looked at warming and animal movement on land but not as much for the ocean. He found the records of fifty years of scientific surveys and bottom-trawl samples. Once assembled, his data set encompassed 128 million individual animals, captured between 1968 and 2011 off both coasts of North America.[51] Then he and his colleagues used temperature records to map "climate velocities," a measurement of both how much the average temperature had changed and what direction it had changed in. The basic question was, How well did the location of fish match their local climate velocity? The answer was: quite well. More than 70 percent of fish were moving in the same direction as the changing ocean they swam in. The results echoed the work of intertidal researchers who had recorded their own creatures marching poleward. At the same time, it raised a thorny problem.

Fisheries management had started to succeed in recovering fish in some places. But, as Pinsky told us, fisheries management was set up on the assumption that the fish would not move from their documented ranges. By the mid-2010s, it was clear this assumption was not correct. The summer flounder fishery, for example, is regulated by the U.S. government under a complicated sharing agreement to ensure that each state lands a certain percentage of the overall catch.[52] In recent years, states have commonly transferred components of their catch quota to neighboring states as the fish change their location through time. One example is North Carolina, where fishing boats may motor out of harbors to catch the roughly 20 percent of the commercial catch of summer flounder that is portioned to that state, only to find that those fish are now more commonly found twenty-five miles to the north.[53] Simply predicting where fish will go is only half of the story because human behavior and decisions also determine what will be fished. Pinsky emphasized these complexities to us, noting that sometimes ecologists might approach conservation thinking about humans as simply imposing on natural systems, but in his view—in an echo of Kawika Winter's idea for saving the world with Hawaiian fishponds—we need to understand that humans are part of the natural world.

Pinsky's work has documented the changing locations of a variety of species and in doing so has inspired a suite of visualizations that enable managers, fishers, politicians, and conservation groups to understand the changing ocean. Summer flounder, Northeast Atlantic mackerel, American lobster—all marching, swimming, moving northward.[54] Of course, many factors can complicate or influence where species move to on a seasonal or annual basis, but Pinsky and colleagues have suggested that over 40 percent of the northward shift could be attributed to climate change.[55] "That's impressively high," Pinsky told Reuters in 2018. "That something as simple as temperature explained a lot of the pattern, given that there's fishing, there's predators, there's prey, de-oxygenation, pollution and changing currents. There's so much going on."[56]

Oceans shift dramatically on their own, outside human influence. From the hyperlocal—the presence or absence of a particular small animal at a particular time—oceanographers might work back to hemisphere-size patterns of wind, water temperature, pressure, and salinity that ebb and flow across the oceans like the tides. The Pacific Ocean cycles between El Niño and La Niña, warm and cool surface temperature patterns, respectively, in the tropics that cause ripple effects outward across the entire ocean basin, leading to dramatic weather changes on land in Southeast Asia, South America, and North America. When an El Niño forms, rain floods the Peruvian desert, the monsoons weaken in India, and seabirds and marine mammals starve in the absence of cold-water prey off the normally chilly coast of San Francisco.[57] A similar but longer-term process called the Pacific Decadal Oscillation, based on sea-surface temperature and pressure in the tropics and northeastern Pacific, swings between "positive" and "negative" phases, with different plankton and copepods blooming in its wake.[58] In the far north, an atmospheric pattern called the Arctic Oscillation swings back and forth every few years, bending or straightening the path of the jet stream and thus influencing the weather far away in the continental United States.[59]

In the Atlantic, from the base of the food chain all the way up to fish and whales, life has followed similar long-term cycles between warm and cold water for millennia. Scientists have tracked the major driver of change, the North Atlantic Oscillation (NAO), since realizing its predictive power in the 1950s.[60] The NAO can be either positive or negative, depending on the average differences in sea level pressure over Iceland and over the Azores. But for something so seemingly simple, the ripple effects are enormous and cover much of the

Northern Hemisphere.[61] When the NAO is positive, with low pressure in the north and high pressure in the central Atlantic, milder weather descends on the Eastern United States, increased rain on Northern Europe and Scandinavia, and drought on Southern and Central Europe. The NAO affects circulation, wind, and temperature in almost every body of water in Europe and eastern North America, meaning it also determines what life is found where.

Although it is a cycle that often takes place over a period of several years, the NAO doesn't follow a perfect symmetrical path. Sometimes it swings up for a few years then dips down for a year then swings up again; sometimes it decreases a little bit each year for four or five years before flipping to its positive state for a while. Through the mid-2010s, scientists had recorded multiple "regime shifts" in the Atlantic as the ocean and the creatures living in it moved around following the changing water conditions. These shifts, taking place over a period of several years or decades, set the biological limits of the North Atlantic. For many years, the NAO offered one avenue for predicting what kind of fish would be found where and how many of them people could catch in a given year.

The NAO is formally defined as the gradient of pressure between the average pressure at sea level as measured in Lisbon, Portugal, and Stykkishólmur, Iceland. To measure its effects and connect it to the rest of the world, scientists needed simultaneous data on wind speed and direction, water temperature, salinity, and biology from multiple places across the Atlantic Ocean and various European seas. And, of course, it took time to understand: decades of observation to watch the gradient flow one way and the effects ripple outward, then watch it flow the other way with more ripples, until it became clear it was a *pattern* and not just random, chaotic change.

To construct a full model of the oceanic oscillations was a project for a globalized world. Perhaps unsurprisingly, then, the origins of understanding oceanographic change became intertwined with the colonial and imperial desires of the early United States. In the 1840s, the U.S. Navy had warships scattered across the North Atlantic Ocean. Naval ships kept measurements of the wind, current, and temperature in their logbooks, and a naval lieutenant from Virginia named Matthew Fontaine Maury realized he could assemble all these logged data points into a single map of the Atlantic Ocean—one of the earliest oceanographic data synthesis efforts. A navigator who had never sailed a particular location before could benefit from the knowledge of thousands of ships that had come before him.

Maury's goal was to improve navigation for the purpose of expanding the slave trade. He pictured a vast, connected American slavery system stretching from the United States to South America. "Maury contributed to the long-standing effort to employ science in the service of empire, a strategy widely recognized by historians of ocean science," science historians Penelope Hardy and Helen Rozwadowski wrote in a commentary in *Oceanography* in 2020.[62] Maury joined the Confederacy in the Civil War and failed in a personal assignment to recruit European supplies for the South. He fled to Mexico after the war, where the short-lived anti-Republican Austrian emperor Maximilian I appointed Maury to found a "New Virginia" and bring Southerners to Mexico. After Benito Juárez recaptured Mexico and executed Maximilian, Maury moved back to Virginia, where he died in 1873.[63] Fifty years later, in a time when white Southerners wanted heroes to celebrate their contributions to the world, they found Maury. A statue went up in his honor on Monument Avenue in Richmond, Virginia, in 1929 and remained standing as Maury became known to more and more Americans as the father of oceanography. (The statue was pulled down in July 2020.)

The roots of scientific oceanography lie here, not in an attempt to benefit "mankind," as Maury's monument association proclaimed, but in the promotion of exploitation and commerce. As Kawika Winter says of fisheries, oceanography, and Western colonial practice more generally: Who does Western science benefit, and who does it leave out? "Maybe you already know something about this," the poet Alexis Pauline Gumbs writes in an essay about North Atlantic right whales. "About how a deadly system doesn't have to seem like it's targeting you directly to kill you consistently."[64]

The North Atlantic right whale is thought to have received its name in the 1700s from whalers who considered it the "true" or "proper" whale to target for hunting.[65] Commercial hunting records for this species go back 1,000 years.[66] Basque fishers hunted the whales first on the European side of the North Atlantic, then on the American side, where the predictable seasonal migrations of right whales along the coast made them a reliable source. Keratin-rich baleen from the whales' mouths was used in a variety of different products, but most valuable was the whale oil. Estimates place the North Atlantic right whale population between 9,000 and 21,000 individuals prior to whaling.[67] Like many of the fish that Europeans crossed the ocean to catch, the population crashed, until there were too few whales left to hunt, decades before they were formally protected by international treaty in 1935.[68] North Atlantic right whales are no longer hunted, but they have never recovered. Today there are an estimated 300–400 individuals.[69] "The precarious status of the northern right whale is due to the fact that commercial whalers specifically targeted and hunted them for the entire period of the slave trade," Gumbs writes. "Is it possible to untangle the consequences of centuries of rapacious greed?"[70]

Michael Moore, a marine biologist at the Woods Hole Oceanographic Institution who is something like the unofficial doctor to the right whales, told us he wanted to be a vet because his father and brother were medical doctors and he wanted to do something different without doing anything *too* different. In veterinary school in the 1980s, Moore said, in crowded basic physiology and pathology lectures, his professors would try to keep them all awake by telling stories about how marine mammals did things differently. The difference appealed to him. "That caught my attention," he told us. "They were the odd ones out, I was an odd person out."

Moore had started following whales as an undergraduate at Cambridge. A project in Newfoundland, Canada, needed research assistants who could help study humpback whales, which got caught in the gear and nets fishers were using in the still-thriving Atlantic cod industry. The whales and the cod wanted to be in the same places in pursuit of a small fish called a capelin, and so the people chasing the cod ended up also meeting the whales. It was Moore's first glimpse of the complicated interactions between marine life, marine conditions, and human behavior. Moore didn't see a place for himself in science because he was interested in applying what he learned, and there were very few people making a career of applied veterinary science on large whales. But the Woods Hole Oceanographic Institution lured him into a PhD program, where he studied the presence of toxic polychlorinated biphenyls (PCBs) in fish. After he completed the PhD program, his interest switched back to North Atlantic right whales and what was happening to them. When a whale would wash up on land and die,

Moore would head for the beach to conduct a necropsy and figure out what had killed it. He noticed that he kept tripping over fishing ropes. In veterinary school, Moore had studied whaling weaponry, including explosive harpoons. He couldn't help but draw a contrast now between the rapid deaths by harpoon that he'd read about and the six months it might take an entangled whale to die as it drags fishing nets or lobster traps through the ocean.

Moore's interest in animal welfare and his background as both a veterinarian and marine scientist led him to work with the National Oceanographic and Atmospheric Administration (NOAA) on cause-of-death investigations, which were required under a variety of new laws to protect marine mammals. But Moore's investigations showed that the cause of death for many of the whales was related to the fishing industry. Fishers grew frustrated with NOAA, stopped reporting carcasses, and started suing to weaken and/or overturn environmental regulations. At the same time, increasingly powerful conservation groups started suing NOAA to strengthen or enforce environmental regulations. "One of the things that's happened to me as an ocean scientist is my increasing familiarity with the legal system," Moore told us.

In the 1950s, researchers realized that each North Atlantic right whale could be identified as an individual. They spent decades constructing a photo library of every known whale, allowing them to try to count them all. The first estimate, reported in 1980, showed how perilously the right whale clung to life on Earth: fewer than 100 whales. Although the number grew slowly through the decade, scientists forecasted in 1991 that the whale was headed for extinction within two centuries. But the number kept growing, and after an especially productive period in the 1990s, there were 486 whales counted in 2010. In a scientific assessment, researchers called the recovery "unexpected," and tied it to a major increase in the presence of the copepod *Calanus finmarchicus* in the Gulf of Maine.[71]

The North Atlantic Ocean is structured by a series of large surface currents that transport heat, salt, nutrients, plankton, and really anything else that is floating along for the ride. The Gulf Stream carries warm water from the Gulf of Mexico, through the Florida Straits, northward along the Florida coastline, and then along the eastern seaboard toward Canada. A portion of this current stretches even further northward, where it is called the North Atlantic Drift, with two arms that split and wrap around each side of Iceland—one arm heading north into the Norwegian Sea and toward the Arctic Ocean, another meandering westward toward the Labrador Sea, between Greenland and Labrador. This oceanic river of warm water moves at about four miles per hour, transporting warm tropical water to the high latitudes.

An equivalent current system moves southward along North Africa, bringing colder Arctic waters toward the Equator. These current systems are mimicked in each ocean basin, with transport of heat and salt away from the Equator toward the poles, and cooler, more nutrient-rich waters moving toward the Equator from the poles, in a gyrelike pattern.[72] But the North Atlantic currents are also places where the surface currents interact directly with a deeper ocean current system. This deeper ocean circulation system—called the meridional overturning circulation because it moves waters across lines of latitude (meridians) and across depths (overturning)—is driven by differences in the density of the water itself. In the far North Atlantic, as warm, salty Gulf Stream water cools, it becomes some of the densest water in the ocean and sinks to thousands of meters below the surface. That water then joins a global circulation system of waters moving at depth due to density differences and eventually upwells back to the surface.

Because of climate change, the waters that sink in this region and join the deep ocean circulation system seem to be slowing down.[73] The characteristic fingerprints of this slowdown are a cooling in the subpolar Atlantic, a warming in the Gulf Stream, and an associated northward shift in the location of the Gulf Stream.[74] In addition, the Gulf Stream appears to be bringing those warm waters closer to shore, bringing heat to the Scotian Shelf and the Gulf of Maine.[75] These changes in the North Atlantic have resulted in some of the fastest warming waters on the planet.[76] While the oceans as a whole warmed by about 0.01°C every year between 1982 and 2013, the Gulf of Maine warmed 0.03°C every year over the same period. The rate of warming in the North Atlantic accelerated after 2004, and for the next decade, it averaged to 0.23°C per year. That's a remarkable upward trend for an area that swings dramatically between cold and warm—a sustained, intense heat wave that "few marine ecosystems have encountered," one group of scientists wrote.[77] By 2010, "the oceanography changed dramatically," Cornell oceanographer Charles Greene said in 2021.[78] There's a new "ecosystem regime" for the region, Greene and his colleagues have concluded, one that is not responding only to the various oceanic oscillations.[79] As in the tide pools and the kelp forests, with higher temperatures comes the changing distribution of species, introducing new organisms that didn't previously thrive in this region.

Commercially valuable fish might make national headlines when they move poleward, but perhaps the greatest indicator of the shift in the Atlantic is in the tiny *Calanus finmarchicus*. A millimeter-scale crustacean called a copepod, *Calanus finmarchicus* is rich in lipids and provides the primary source of nutrition for commercial fish like herring, haddock, and cod, as well as calves of the North Atlantic right whale.[80] In general, the colder the water, the more *Calanus finmarchicus*. The copepods could be found traditionally in small numbers as far south as Cape Hatteras, North Carolina, increasing to massive swarms in the Labrador Sea.[81] Now, after powering a North Atlantic right whale boom in the Gulf of Maine in the 1990s, the copepods have found conditions more to their liking on the other side of Newfoundland.[82] All the life that relies on them has tried to follow. "The North Atlantic right whale used to be very reliable in its preference for feeding on high densities of copepods in the Bay of Fundy," Moore told us. "Around 2010, the incursion of the Gulf Stream waters into the Gulf of Maine made that a less optimal place for the copepods to do their thing, and the whales said, 'To hell with it, nothing here worth eating.' They showed up in the Gulf of St. Lawrence."

In his 2021 book *We Are All Whalers,* Moore wrote, "It is patently obvious that relying on the current laws and regulations of the United States and Canada has yet to ensure conservation of the North Atlantic right whale species or prevent severe, prolonged suffering for individual animals."[83] When the right whales suddenly quit the Gulf of Maine and moved north into areas in Canadian waters they had not typically occupied, the result was a rapid increase in deaths due to ship strikes and entanglement in fishing gear. For the last decade the whales have also generally had fewer calves, possibly because dragging ropes and gear when they survive entanglements makes it harder to reproduce and raise calves.[84] From a 2010 high of just under 500, the population has dropped dramatically again to fewer than 360. An annual report card on the right whale's status provided by the North Atlantic Right Whale Consortium reads like a student struggling, and failing, to recover from difficult coursework.[85] Scientists, fishers, and conservationists alike watch the whale calving season with great attention, celebrating each individual birth and mourning each death. The NOAA website highlights "meet the mothers and the calves" of each year, with a name and number and story for each mother and calf pair, for example:

Grand Teton was seen with her new calf off the coast of Florida on January 11, 2021. Grand Teton is approximately 40 years old, and this is her eighth calf. Grand Teton's 2010 calf Mayport (#4094) was last seen entangled during the summer of 2017. She had previously calved in 2016, but that calf has not been seen since 2017 either.[86]

The number of dead whales washing up on the beach has increased to the point that NOAA declared an "unusual mortality event" every year between 2017 and 2022. In each case of a confirmed whale mortality with a carcass, the whale identification, cause of death, and location is noted.[87]

In addition to observing whales from ships, airplanes, and the shore and tracking changes in plankton using nets, scientists simply listen to the lives and migratory patterns of the right whales. Passive acoustic monitoring—the continuous measurement of whale calls via gliders, buoys, and bottom-mounted recorders—has documented the whales as they moved more frequently into the busy shipping and fishing area of the Gulf of St. Lawrence. "From our acoustic monitoring and other tracking efforts, we know that North Atlantic right whales are a true coastal species, directly overlapping with the densest areas of human activity," says NOAA zoologist Sofie Van Parijs, who started NOAA's Northeast passive acoustic research group in 2006.[88] "This puts them at greater risk than other whale species like blue whales, which are more of a deep-water species."

Prior to 2010, scientists, managers, and fishers felt that the right whale could be a conservation success story. Vessel speed reductions in key whale migration routes and efforts on the part of lobster and gill net fishers had kept the mortality rate of right whales below an acceptable threshold.[89] Then, geographic shifts in the food web caused these previously "successful" management steps to fall apart. In a conversation with us, Brady O'Donnell, a communications and legislative affairs officer at the U.S. Marine Mammal Commission, noted that "people now recognize that this is a conservation problem with a species that is wrapped up in climate change." And now the effects of climate change on the whale populations may take an unexpected turn: federal governments are considering plans for offshore wind operations as a method to reduce fossil fuel emissions, and there is growing concern that the placement of offshore wind turbines will affect both the fishing industry and the right whale populations. Passive acoustic monitoring will be used to understand the overlapping ocean territories of right whale migrations and potential wind farm locations.[90] As people continue to explore multiple uses and needs for the ocean, the migratory routes of whales—termed blue corridors and whale superhighways—are coming into focus as national and international conservation and management challenges.[91]

Moore told us that maybe the biggest challenge of all is simply alerting the public to the horrific nature of death by entanglements. Much like coral researchers, whale scientists have had to grieve the loss of the subjects they study even as they grow increasingly frustrated at decades of inaction that might save them. Moore suggested that, if people knew what it was like to watch a whale die, slowly, over six months, weighed down by the increasingly heavy burden of fishing ropes, then they might be convinced to buy more ethically sourced seafood or to demand the crab and lobster industries switch to ropeless gear, and the right whale might be saved. He writes about how the whale's problems tie specifically to what people buy: our demand for crab and lobster, manufactured goods shipped from around the world, and crude oil and its derivatives is harming the whales. "In spite of all that, in spite of the individual understanding of levels of trauma and the six months it takes to die from trauma, there's been a blindness or out of sight out of mind problem in terms of engaging the consumer," Moore told us. "If an entangled right whale habitat was the streets of a major city, [if people saw] individuals of those species dragging lobster traps down the street while people were going to work each day, they wouldn't buy lobsters anymore.... How do we deal with that communication gap? It's something that keeps me awake at night."

The North Atlantic right whale needs a superhighway that moves from the eastern seaboard into the Arctic, and likewise blue whales, gray whales, and humpback whales are in the midst of a traffic crisis on the North American West Coast. Overlapping migration routes that put the whales in direct contact with fishing gear and fast-moving vessels are a threat there, too. Unlike the right whales of the Atlantic, the large Pacific whales have, in many cases, seemed on the path to recovery from whaling pressure. The Eastern Pacific population of the gray whale was removed from the Endangered Species List in 1994, and some populations of humpback whales were delisted in recent years as well. Yet they too have experienced challenges in recent years from global climate change and the many human demands made of the ocean. Whales in the California Current ecosystem face threats from fishing gear entanglement, vessel strikes, noise, water quality, and marine debris.[92] Climate change further complicates the picture, interacting with the other threats to whales by changing oceanographic and ecological patterns and increasing potential sources of disease.

A few years after the Atlantic Ocean seemingly switched to a new regime, the Pacific Ocean followed. In 2015, forecasters noticed an El Niño pattern forming in the tropical Pacific, but where cold water normally wraps around the warm pattern, this time it was just warm everywhere, across essentially the entire ocean basin. In 2016, at an ocean sciences conference in San Francisco, a long-time coastal scientist labeled the conditions in the Pacific a "no-analog ecosystem."[93] The changes taking place were so fast, so widespread, and so significant that they had led to "unprecedented redistributions, mortality events, toxicity, and socioeconomic impacts." The Pacific had entered a new state unlike any seen, or even measured, in the past.

The heat wave that decimated kelp forests on the U.S. West Coast between 2014 and 2016 also led to a massive harmful algal bloom. In 2015, at the peak of the heating, a warm-water-loving diatom called *Pseudo-nitzschia* rapidly expanded across much of the eastern Pacific.[94] *Pseudo-nitzschia* produces a deadly neurotoxin called domoic acid, and as the diatoms spread, so did the toxin. All summer, up and down the West Coast, beachgoers reported cases of marine mammals acting strangely as the poison took hold in their brains. Sea otters, seals, and sea lions suffered seizures or tremors, or walked, disoriented, into cities. Elsewhere in the food web, filter-feeding shellfish and animals like crabs accumulated domoic acid in their tissues, forcing fisheries managers to delay the opening of commercial crab season and prohibit clam and mussel harvests. In San Francisco, where the crab season typically opens in late fall and Dungeness crab is a traditional Christmas dinner staple, the fishery didn't open until March of the following year.[95] Crab pots sat in giant unused piles in port towns from Santa Barbara to the Oregon border, and the fishery estimated the costs of lost sales at over $40 million.

When the fishery did open in March, it was to fanfare and genuine delight. But now it was a different time of year for fishing. Instead of the traditional winter ocean, crab fishers headed out into the middle of the spring whale migration. And the warm water didn't just lead to the domoic acid poisoning—it came with significant movement of marine species, mirroring the heat wave in the North Atlantic just a few years earlier. The krill and small fish prey species favored by gray, humpback, and blue whales prefer cold water so they swarmed closer to shore in search of cooler, more nutrient-rich refuges.[96] The whales followed, bringing them directly into the path of the crab boats and their gear. When the crab season reopened, a record high number of whales of all species became entangled and died. In most cases where gear could be identified, it appeared to be associated with the Dungeness crab fishery. "In many of the more well-documented cases, climate extremes amplify human-wildlife conflict," NOAA fisheries biologist Jameal Samhouri and colleagues wrote in 2021.[97]

In November 2018, the Pacific Coast Federation of Fishermen's Associations, a trade group that represents Dungeness crab fishers, filed the first industry-led lawsuit against fossil fuel companies, seeking repayment for damages from the fishery closure. The lawsuit alleges that the fossil fuel companies had known for decades that use of their products could be "catastrophic" and that "only a narrow window existed" for action before consequences would be irreversible. In its introduction, the lawsuit states, "the crab fishing industry brings this action to force the parties responsible for this severe disruption to fishing opportunity, and the consequent impacts on fishing families, to bear the costs of their conduct."[98] The lawsuit joins a sea of others brought by state and local governments against fossil fuel companies for damages associated with climate change.[99]

The California crab fishing community and the Gulf of Maine and Gulf of St. Lawrence lobster fishing communities find themselves at a similar crossroads now: how to protect the whales from entanglements while maintaining coastal economies and sustainable fisheries. Many marine conservation and engineering groups see the development of ropeless gear, or gear that uses acoustic triggers to pop up to the surface after deployment, as a logical solution to the entanglement problem. Michael Moore told us that he had heard suggestions of acoustic gear triggers as far back as 1979. But so far, they have not been widely adopted. "That was 1979," Moore says. "It's a long time ago. Change didn't come nearly as fast as it should."

Some West Coast crab fishers have been willing to test ropeless gear, and others have balked at the cost of replacing an entire statewide fishing fleet with what they contend are expensive new crab pots that do not work as well. The fishing communities have pushed back against what they see as a "top down" approach led by conservationists who do not fish and that require new technologies that the fishers haven't been involved in developing. "The objective would be to get the fishermen to be the majority of the designers and developers," said California crab fisher Dick Ogg, who has tried to work with scientists and California policymakers to make the fishery more sustainable.[100]

In California, fishers, managers, and scientists have endeavored to prevent another year like 2015. Voluntary vessel speed reductions are in place for shipping lanes at the Port of Los Angeles, Port of Long Beach, and outside San Francisco Bay to try to reduce collisions with whales in busy shipping lanes—although so far, most ships have not followed them.[101] (Voluntary speed reductions also reduce the emissions of greenhouse gases and other pollutants.) NOAA, the Coast Guard, and multiple national marine sanctuaries in California have created incentives and awards programs to try to encourage ships to reduce their speed, and they also track annual observations of ship strikes. There were seventy recorded observations of ship strikes on "large whales" (gray, minke, blue, humpback and fin whales) from 2007 to 2020, and research suggests that this is probably a small portion of the actual number of strikes.[102]

In spring 2022, the state of California announced that it would close the crab fishing season early after two humpback whales got caught in crab gear off the coast of central California. This has been the generalized approach in lieu of the adoption of new gear technology since 2017—limiting the overlap in the commercial crab fishing season and the seasonal whale migration. In a note explaining the decision, California Fish and Wildlife director Charlton Bonham said that the state was trying to "strike the right balance" between whales and fishing.[103] In a broader sense, the idea of finding balance is relatively new in a government tradition that has, for most of its existence, viewed the ocean as limitless. In a 2022 article in the journal *Marine Policy*, several scientists, including shellfish disease researcher Priya Shukla, wrote that "the contemporary seascape of whale mortality necessitates a broader response based in systems thinking."[104] Climate change only makes the attempts to find balance more difficult. "The solutions to our problems are tractable given the whale," Moore says. "Certainly we're not going to change climate change driven factors in time. But having said that, the reduction in vessel trauma and entanglement trauma is eminently doable if there's a will to do it. It comes down to economics, politics, and culture. Those are the issues."

The heat wave and harmful algal bloom events aren't even the first time that West Coast crab fishers in particular ran into the effects of climate change on the ecosystem they rely on. A well-documented low-oxygen mass of water has developed nearly every summer for the last two decades off the Oregon coast, hovering above the sea floor and suffocating ecosystems poorly adapted to such conditions. Fishers pull up crab pots full of dead crabs during this now annual "season" of low-oxygen waters, and news articles have shown thousands of dead crabs and fish washing up along the shoreline, filling tide pools and littering the beaches. These low-oxygen waters, also called hypoxic—containing too little oxygen to support most ocean life—"suggest a fundamental shift in ocean conditions off the Pacific Northwest coast" since 2002, notes the Partnership for Interdisciplinary Studies of the Coastal Ocean (PISCO), a leading research group investigating the phenomenon. While it is normal to find lower-oxygen waters deeper offshore (associated with the long and deep decay of organic matter drifting from the surface), it was once considered unusual for such waters to come into water depths of less than 165 feet. Despite decades of scientific monitoring, fishing, and observations along the Oregon coast, the inshore hypoxia season hadn't been reported before 2002.[105] Now that it is a high-impact annual phenomenon, it has been a research priority, and scientists have developed highly skilled models that can examine current oceanographic conditions and predict, on a six-month timescale, when and how dramatic the hypoxic zone may be.[106]

In 2021, the hypoxic zone started earlier and hung on for even longer than normal. The north winds that churn up the ocean along this shoreline began early that year, bringing low oxygen waters up to shallow depths, and kept going strong throughout the summer and into the fall. "Back in April [2021], we predicted it would be a bad hypoxia year because of the weather, and it's turned out to be a really bad year," said Francis Chan, a lead PISCO scientist at Oregon State University, in September 2021. "Oxygen levels got very low, very early, and the worst is not over. On Aug. 31, it was as close to zero as we'd seen this year. Now we're in September, at a time when we thought oxygen would have been rising for a while, and it's just this endless summer, but in a bad way."[107] In 2022, fishers and scientists on the Oregon coast partnered to begin to fill in the gaps in understanding where and when hypoxia occurs. Chan and colleagues developed a small oxygen sensor that can be attached directly to the crab pots. Fishers are now the first to know when waters are trending toward hypoxia and exactly where.[108]

The promise of a vast and limitless ocean always meant there would be enough space for everyone to do everything. Picture our human ancestors 100,000 years ago, picking through the tide pool like a twice-a-day resort buffet; what an endless gift that ocean promised. Rutgers fisheries biologist Malin Pinsky talked to us about the sheer abundance of the past ocean, even into the recent past as recorded in memories, photographs, trawl surveys, and written histories. "Hearing the historical descriptions of how many cod there were in Cape Cod Bay, or how many oysters were in the Chesapeake—the ocean seemed so vast, so hard to change," he told us.

The ocean of a century ago was so radically different that it can be difficult now to know how much has been lost to fishing methods predicated on a limitless abundance. But scientists around the world have found a creative way to try to estimate. As industrialized fishing arrived in new places, surveying trawl ships would go out and drag nets along the bottom to prospect for fishing grounds. They'd land and report their catch to interested commercial fishing operations. These catch records, many of which have just been digitized, offer researchers a way to look back in time. By recreating such trawls today (critically, using the same technology), researchers can rough out something like a classic transect survey. Instead of feeling through the tide pool by the light of a lantern in search of old bolts, these scientists focus on understanding late nineteenth- and early twentieth-century fishing technology.

In 2015, South African National Biodiversity Institute marine scientist Jock Currie and colleagues recreated a trawl design used by a government survey ship in 1903 and 1904 to explore the Agulhas Bank, the vast, shallow shelf extending to the south of South Africa. Unlike many of the other places that had been studied, Currie later wrote, there had been essentially no fishing at all on the Agulhas Bank fishery before the government survey ship *Pieter Faure* showed up in 1897. Using contemporary records from the early 1900s, Currie and his colleagues built a replica otter trawl net woven of Manila hemp in the style of the time. They towed it from a more modern trawl barge at a historically accurate 2.5 knots—roughly a human walking pace. After recreating several dozen surveys in three different areas of the bank, they examined their catch and documented a "transformed" assemblage of fish, with a shift of species that previously indicated reef environments toward species that prefer sandy or muddy habitats.[109] One large predatory fish, the kob, was among the most common fish caught in the 1900s and had all but disappeared in 2015. "I think the cliché 'If you don't know where you've come from, you don't know where you're going' is very relevant in fisheries or biodiversity management," Currie said in a news interview. "Trying to estimate how many fish are left in the ocean and how much fishing is sustainable or not inherently requires the knowledge of how many fish there were before we started fishing."[110]

The field of conservation paleobiology extends these historical records of fish abundance and assemblage even further back in time. In one study, scientists looked at the predation scars—marks left on shells by crabs attempting to crush and eat the creature within—in modern snail shells in Southern California and compared them to the number and kind of marks found on fossilized snail shells from the same area roughly 80,000–120,000 years ago. They discovered that it is much easier to be a snail today. The live snails they observed grew larger and had fewer scars. It was clear from the data, the researchers wrote, that the number of crab attacks on snails has decreased dramatically compared to 80,000 years ago—which, they infer, means there are far less crabs around to attack snails today.[111] Researchers have also drawn conclusions about shifting baselines in Southern California and beyond by using archaeological and fossil records to compare intertidal shellfish assemblages pre- and post-settler colonization.[112]

Every attempt scientists have made to reconstruct fisheries decades or even centuries in the past paints a picture of an ocean changed in dramatic ways. So far, much of that change has been caused by overfishing. Now warming and acidification are rapidly making the picture even more complex. Climate change fundamentally alters the productivity of the ocean in ways that will feel personal to fisheries around the world. As those changes accelerate, the call to protect areas of the ocean that can allow environments to rebound, rebuild, and replenish grows louder.

Angelo Villagomez was born on Guam, the neighboring island of his home on Saipan, in the Northern Mariana Islands. His earliest memory of the ocean, he told us, was going to Saipan's Obyan Beach, a classic Pacific Island stretch of white sand and glittering coral. He was three or four years old and his dad and uncles hoisted him into a small dinghy to travel out over the reef to deeper water. "I can remember being on that boat, being on the edge, looking down, and seeing a rainbow of colors and a cacophony of fish swimming around," he said. "That no longer exists. The corals for the most part have been utterly destroyed. It's gone. It's an ecosystem that no longer exists. I don't have children but if my nieces and nephews were to get on a boat and do the exact same thing they would see rubble."

Villagomez moved to Massachusetts when he was young. He intended to be a scientist, he told us, but after graduating college found himself drawn much more to conservation activism than academia. He has always had an interest in pragmatism, he told us, which led him to the Pew Charitable Trusts and then the Center for American Progress working as a scientific adviser on behalf of major marine policies. At Pew Charitable Trusts, he worked as an adviser on international shark trade policy and the creation and expansion of U.S. national marine monuments, including Papahānaumokuākea Marine National Monument in the northwestern Hawaiian Islands and the Mariana Trench Marine National Monument in the deep sea 500 miles north of his home village.

Villagomez also worked with the International Union for the Conservation of Nature (IUCN) on a resolution calling for protecting 30 percent of the world's oceans by 2030, a policy often called 30×30. There are questions about all marine protected areas, Villagomez acknowledges. Square mile and percentage targets make sense to people but don't actually reflect real conservation—everything depends on where those lines are drawn. Coral reefs and kelp forests worldwide testify to the ineffectiveness of protection against the heat waves and acidification wrought by human carbon emissions. And to be effective, protections for any protected area have to involve the people they affect.

Even in 1883, Thomas Huxley warned against Western governments imposing rules that would tell a fisher he couldn't fish where his father had before him—a constant challenge today, Villagomez says, because traditionally white-led Western conservation organizations push protected areas that prevent fishing or extraction in the global south. He also said, however, that he thinks some conservation organizations are growing and listening. And the challenges haven't dissuaded him from the main point: marine protected areas are important and critical to making the future better.

"My dad's godfather, he was a native Chamorro living in Saipan," Villagomez said. "He has stories from the 1930s and '40s. He would tell stories of tuna inside the lagoon, an outrageous idea today. There were so many fish you almost didn't need to fish. You would go out in a boat and they would jump into a boat. I can't even envision what that looks like. And I've experienced that in my own lifetime. Thirty, forty years ago there was more life in the ocean. More sound, more life, more things living there. Today when you go in that same ocean, you see fewer fish, smaller fish. That's extremely sad. There are kids on Saipan who are learning to spearfish—17-, 18-, 19-year-old kids. They catch fish, and you see the fish they hold up and they're so proud. And fish that size—we would have completely ignored the fish they're catching."

Like Kawika Winter, Villagomez reminded us that the world needs abundance thinking in a time of apparent scarcity. Villagomez says that it's clear we can't think of the ocean as the unconstrained Wild West anymore. "If we don't realize the ocean frontier is closed, and completely change our relationship with the ocean, we're in trouble," Villagomez says. "The United States invented the concept of national parks. We need a similar change in worldview if future generations are going to know the ocean the way that past generations did."

It became apparent to us that like the canoe that Winter and Villagomez described, the ocean is an island. We influence every part of it—but we can employ that influence to make something positive. To accomplish this, we'll have to look not only at our coastlines but also continue into the open ocean to understand the past, and the future, of how we know the ocean.

CHAPTER 6

THE OPEN OCEAN

> *In conservation, we are taught that practices and approaches that were successful should be what we apply to different regions, places, and communities. This tends to ignore that every community holds a different set of values and relationships with their environments.*
>
> —Jessica Hernandez, in Fresh Banana Leaves

On a sunny day in September 2018, Lehua Kamalu navigated a seventy-two-foot traditional Hawaiian canoe called *Hikianalia* under the Golden Gate Bridge and into San Francisco Bay from the open Pacific. Kamalu and her crew of thirteen had just sailed across 2,800 miles of open ocean, from Hawai'i to San Francisco, using only wind and solar power.[1] Two decades earlier, *Hikianalia*'s sister ship *Hōkūle'a* visited the shores of Northern California on a worldwide voyage to demonstrate the skill of ancient Polynesian ocean navigation.[2] The *Hikianalia* arrived in 2018 with a similar connection to the past and also with a conservation message for the future.

Members of the Muwekma Ohlone tribe greeted the *Hikianalia* as it arrived in Aquatic Park in San Francisco. The boat docked for public tours, where visitors picked carefully over the deck, gawking at the giant steering paddle, and asked questions about ocean plastic pollution and the fishing along the way. But Kamalu and Nainoa Thompson, a master navigator and president of the Polynesian Voyaging Society, stepped out of the canoe and onto the stage to address an audience of dignitaries at the Global Climate Action Summit hosted by California's governor.[3] Their message, like Kawika Winter's message from the Hawaiian fishponds, was that, in a world of pressing environmental challenges, the navigators have wisdom to share. Speaking first, Thompson emphasized the size of the ocean. The triangle of Polynesian culture, stretching from New Zealand to Hawai'i to Rapa Nui, encompasses an area three times the size of the continental United States. "I come," Thompson said, "from the biggest nation on Earth—10 million square miles of mostly ocean . . . from an ocean country." Kamalu then added, "We were sailing across the ocean. But that's what we do. Through that practice we learn to be connected to the ocean, to everything that's in it. To the fish that feed us, to the wind that powers our sails, to the signs that guide us from one place to another. . . . I have learned what it means to take action each day to live a healthy and thriving life with the people and place around you."

The settling of the islands of the South Pacific is perhaps the greatest feat of navigation in human history. The people who voyaged between these remote archipelagos likely had a closer knowledge of the open ocean than any people before or after, and after two centuries of genocidal cultural suppression, the last fifty years have seen an oceanwide renaissance in efforts to reclaim that ancestral relationship. The more the new generation of voyagers and scientists learns today, the clearer it is just how sophisticated a map of the ocean Pacific Islanders created in the thousands of years prior to European contact.

At the very southeastern edge of Asia, the land bends away from Thailand and Malaysia in a long arc. As you travel southeast, tracing the edge of the Sunda Plate and deep Java Trench, the continent fractures into increasingly separated islands, from the large Indonesian mainland of Sumatra to the smaller Java, to a series of islands called the Lesser Sundas. On one of the Lesser Sunda islands, called Flores, in a rock layer studded with the remains of Komodo dragons and long-extinct elephant-like Stegodons, archaeologists have unearthed stone flakes and choppers dating to between 800,000 and 900,000 years ago. These are human tools, but they are not the work of *Homo sapiens*, who at that time hadn't made it out of Africa. Instead, if the age range is correct, these tools were made by *Homo erectus*.[4] What is fascinating, a research team reported in the journal *Nature* in 1998, is that Flores is separated from the mainland Sunda shelf by three deepwater straits. And based on past sea levels, at no point in that time window were those straits any narrower than twelve miles. The British biologist Alfred Russel Wallace, traveling through the Sundas in the 1850s, drew a line through one of those straits as a clear biogeographic divider separating Australian and Asian species.[5] (The same Thomas Huxley who called the cod fishery inexhaustible named this "the Wallace Line.") Somehow, nearly 1 million years ago, people ended up on the other side of the line. "Either there must have been a land bridge linking Flores with mainland Southeast Asia in the Early Pleistocene epoch," the researchers wrote, "or *Homo erectus* in this region had the capacity to make water crossings." There is no evidence, the researchers add, for a land bridge.

Even as *Homo sapiens* later moved in and replaced *Homo erectus*, the islands off southeast Asia remained the center of early maritime technology.[6] Women and men launching from one of these islands 50,000 to 60,000 years ago crossed the open sea to settle Australia. The colonization of Australia has long marked a definitive boundary on human voyaging: there has never, in human history, been any point where it was possible to get to Australia without a boat. Therefore, the oldest human evidence in Australia also is the oldest evidence for the capacity of people to cross an ocean.[7]

Australian evidence is particularly useful because it is so clear-cut in a field that can be challenging. In the critical window when people spread out of Africa and clearly must have been learning about traveling on the ocean, global sea levels were lower and coastlines were in vastly different places than they are now.[8] The nearshore places where these early humans would have experimented with boats and other coastal technology are now buried deep underwater—leaving terrestrial-minded researchers to undervalue the importance of the ocean in human history.[9] "The great handicap is that a boat's wake is ephemeral," writes anthropologist Alice Beck Kehoe.[10]

Although there's still considerable disagreement about exactly where the voyages to Australia started and exactly how they were made, there's consensus that this was a remarkable accomplishment for early *Homo sapiens*. "The journey of the first humans to Australia is one of the most important events in history," Yuval Noah Harari writes in his best-selling human history *Sapiens*. "It was the first time any human had managed to leave the Afro-Asian ecological system— indeed, the first time any large terrestrial mammal had managed to cross from Afro-Asia to Australia."[11]

From that point on, human ocean crossings begin to happen even more rapidly. There is evidence that between 25,000 and 40,000 years ago, people moved from Indonesia into Papua New Guinea and the islands to the east, now often grouped under the name Melanesia. Around the same time, people traveled northeast to the Philippines and Taiwan, and from the Asian mainland to settle the Ryukyu Islands and Japan.[12] Relatively speaking, not long after—at least 20,000 years ago—people in East Asia moved northward to follow coastal routes into and throughout the Americas.[13] More recently—likely around 3,000 to 4,000 years ago, Melanesian voyagers began crossing thousands of miles of open ocean to settle Samoa.[14] Researchers suggest that these voyages took place because of limited resources and a growing population that sought new lands.[15] Then, after a well-studied pause in ocean exploration (called the Long Pause in archeological studies), Pacific voyagers continued oceanward toward Central and Eastern Polynesia and Micronesia, settling there around 1,000 years ago.[16]

Reconstructing the exact histories and timing of these voyages has been a challenge and has included at times conflicting archeological and genetic evidence about when and how people arrived, settled, and lived. Polynesian oral traditions can reliably tell you where their ancestors came from, and from genealogies, estimations can be made of how long ago they arrived. These oral histories can be matched with indirect evidence of human occupation and contact. Some things travel with humans wherever they go, and so, much like how researchers infer boats from the settlement of Australia, they infer the settling of a particular island from the presence of the things that reliably travel with humans. For example, distinctive red pottery appears in Melanesia around 1500 BCE and then in Tonga and Samoa within 500 years.[17]

Radiocarbon dating of fragments of wood, plants, eggshells, bone, and marine shells provides date ranges of Polynesian voyaging: 1025–1120 CE, for example, for the Society Islands, an archipelago in today's French Polynesia that includes Tahiti. After a few hundred years, possibly aided by the invention of stronger, new, double-hulled canoes, another pulse of immigrants set off from Polynesia for Aotearoa New Zealand, Rapa Nui, and Hawai'i.[18] Rather than a series of unidirectional pushes, Beck Kehoe describes Pacific voyaging as moving in "every direction within the huge Polynesian triangle," which is defined by the Solomon Islands to the west, Hawai'i to the north, New Zealand to the south, and the American continents to the east.[19] There is even evidence, based on shared boat construction techniques, that Polynesian voyagers met and worked with the Chumash people in Southern California.[20]

Pacific Islander exploration was not confined to the discovery of new lands. Ample evidence exists for "sustained contact, in some cases over centuries" between disparate island communities, Cambridge anthropologist Nicholas Thomas writes in the 2012 book *Islanders*. "The lives of the people of Oceania were already complex, unstable, and political, before the arrival of the Europeans," Thomas continues. "For as long as there had been islanders in the Pacific, they had traveled and traded extensively. They had regularly taken great risks to seek out new lands. They had long dealings with people who were more or less unlike themselves."[21]

On a map, the Hawaiian Islands are a dot in what looks like the exact center of the Pacific Ocean. An archipelago of eight major islands and numerous smaller atolls, Hawai'i lies 2,900 miles west of Cabo San Lucas, Mexico; 5,300 miles east of Taiwan; and 2,600 miles south of Kodiak, Alaska. But Polynesian navigators made the voyage there with confidence 1,000 years ago. Although Native Hawaiians developed a culture and religious beliefs distinct from other Polynesian islands, there are strong shared connections between all of the Polynesian islands in language, history, and culture—showing how the skills of open-ocean navigators kept these communities intertwined despite being separated by thousands of miles of blue ocean.[22] Native Hawaiians call the route between Tahiti and Hawai'i Kealaikahiki—literally, the "road to Tahiti"—but which modern Hawaiians also translate as "ancient sea road" or "heritage corridor."[23]

Seventy percent of the Earth's surface is ocean, yet the coastal habitats we know well, including tide pools, coral reefs, seagrasses, and kelp forests, make up only a small part of this watery world. The rest may look like a consistent blue desert, stretching unbroken from horizon to horizon and defined by what it is missing. Polynesian voyagers, however, mapped the blue expanse, made sense of it, and turned it into something modern navigators can compare to a highway with on-ramps and off-ramps.[24] The Tahitians, historian Anne Salmond writes, saw the Pacific Ocean as "a flat plane ... crossed by sea-paths between clusters of the known islands."[25]

Early Pacific navigators understood some of the same phenomena that we seek to learn more about today: the interactions between the tides, waves, winds, and currents; how currents consistently direct ocean water from one location to another; the shifts in winds and currents with the seasons and storms; and the meaning of the migration of birds or a concentration of fish or marine mammals. Navigators would have understood the energy of the trade winds north and south of the Equator, and the weakened winds near the Equator (now referred to as the doldrums, where sailors may drift for weeks waiting for a breeze to pick up). There's also much they would have known that, from a modern perspective, is hard even to summarize. Polynesian societies kept records through oral traditions, with navigators memorizing the voyaging directions from island to island—even dozens of islands they'd never visited—in chants. A navigator would also have had to understand specific cultural metaphors that helped interpret signs at sea. Salmond suggests that training and practice then made much of the navigation instinctive. A master navigator and their canoe worked together as one, and the navigator responded to the feel of the ocean against the hull. "Over long years of voyaging which began with an apprenticeship to an older kinsman, the navigator learned to read the sea, stars and winds, until this knowledge became reflexive and embodied," she writes in a 2005 essay.[26]

As familiar as most people are with it today, the cartography system used by eighteen- and nineteenth-century Western explorers is nonintuitive and abstract; it removes the navigator from the perspective.[27] European sailing maps artificially flattened the globe into a grid of parallel lines with a constant orientation in which the north is always up, which made it easier to calculate sailing angles. By constant measurement and math, the Europeans could fix their location anywhere on their rectangular map of the seas, and anyone with the same tools could later follow their path. The charts were "computational devices in which the knowledge of generations of voyagers was embedded," Salmond writes, and the genius of the system, but also the trade-off, was that it made reading the ocean itself inessential. Experienced sailors might acquire something like an "embodied knowledge" of the sea, but that wasn't knowledge they needed to direct the ship.[28]

In the Polynesian tradition, in contrast, "the geographic centre of navigational orientation was inevitably the navigator, and the *pahi*, the voyaging canoe (in Society Islander terms), which was imagined as fixed, surrounded by an animate world of ocean, sea life, wind, current, sun, stars, planets, and ultimately islands," Danish literature scholars Lars Eckstein and Anja Schwarz wrote in 2019. "Wayfinding in this system crucially depended on precise information about the situational bearing of target islands, to be constantly reconfigured in the process of voyaging by closely observing the stars at night, the course of the sun in daytime, and the directions of wind and swell, by observing the wake for current drift and leeway and a range of other factors."[29]

Eckstein and Schwarz take particular interest in a moment when European and Polynesian navigators first attempted to communicate geography to each other across this cultural divide. By the late 1700s, French, Spanish, and English ships had visited several Pacific Islands and anchored for months in places like Tahiti and Hawai'i. Navigators from the islands soon joined European ships as they journeyed onward. A Tahitian named Ahutoru sailed to France with Louis-Antoine de Bougainville in 1768; two more Polynesians piloted Spanish ships in the 1770s, with one sailing to Lima and back.[30] But perhaps the best known of the Polynesian master navigators contemporary with the age of European exploration is the Tahitian diplomat, religious leader, and royal consort Tupaia. An exile from Ra'iātea, at the time the center of Polynesian navigation tradition, Tupaia became friendly with the English and asked to sail with James Cook as the English departed Tahiti in 1769.

Tupaia didn't just help the English navigate the Pacific. He wanted his knowledge documented, and he worked closely with the ship's crew to do so. He painted scenes of Tahitian life and customs, and advised English illustrators on place names and geographic details. And as the ship sailed through Polynesia and south toward New Zealand, Tupaia sat down with the artists to make a navigational chart of everything he knew. None of his maps are known to survive, but three roughly contemporary copies made by English cartographers reflect Tupaia's work, with added layers of interpretation. Tupaia's map is "one of the most famous and enigmatic artifacts to emerge from the early encounters between Europeans and Pacific Islanders," Eckstein and Schwarz write. The confusing disagreements between these archived maps, each labeled with something like "according to the notions of the inhabitants of O Taheitee and the neighboring isles chiefly collected from the accounts of Tupaia," has made the actual information shared by the master navigator an enduring mystery.

What emerges from scholarly investigation, however, is a consensus understanding of the South Seas not as a vast and limitless ocean punctuated by tiny dots of land but what the late Fijian essayist Epeli Hau'ofa described as a known and charted "sea of islands." The European view, like the Mercator projections that undergirded their charts, saw the islands from the perspective of a fixed point centered far away. In Tupaia's world, the ocean connected rather than isolated the Pacific Islands. Part of the project to reclaim voyaging today, Hau'ofa and others argue, is to reclaim that non-Western view.

"If we look at the myths, legends, and oral traditions, indeed the cosmologies of the peoples of Oceania, it becomes evident that they did not conceive of their world in such microscopic proportions," Hau'ofa wrote. "There is a world of difference between viewing the Pacific as 'islands in a far sea' and as a 'sea of islands.' The first emphasizes dry surfaces in a vast ocean far from the centres of power. Focusing in this way stresses the smallness and remoteness of the islands. The second is a more holistic perspective in which things are seen in the totality of their relationships."[31]

With their maps emphasizing distance and negative space, it became difficult for subsequent European historians to explain the settlement of the Polynesian Islands. The Māori told European settlers and missionaries their ancestors had come by canoe in waves of migration from a place they called Hawaiki, and they shared details: "why and how they left Hawaiki, the names of the voyaging canoes and principal crew members, the troubles experienced en route, and where exactly the migrants landed along the coast of Aotearoa," anthropologist Ben Finney wrote in 1991.[32] Yet into the 1960s, some white anthropologists believed that the settlement must have been accidental, and the stories told about the open-ocean voyages were convenient myths meant to reinforce land claims in the present.

Hau'ofa jokingly dismisses the idea of accidental voyaging, writing in his famous 1993 essay *Our Sea of Islands*: "Only blind landlubbers would say that settlements like these, as well as those in New Zealand and Hawai'i, were made through accidental voyages by people who got blown off course—presumably while they were out fishing with their wives, children, pigs, dogs, and food-plant seedlings during a hurricane."[33]

Others took the challenge more literally. In 1973, Finney, a white American who had grown up in a U.S. Navy family in San Diego, and Herb Kawainui Kāne, a Native Hawaiian historian, artist, and Navy veteran who had grown up between Hawai'i and Wisconsin, started a project to prove that the settlement voyage was intentional by building a double-hulled canoe and sailing it from Hawai'i to Tahiti using only traditional navigation. A voyaging canoe had not traveled between Tahiti and Hawai'i in 600 years when Kāne first considered reconnecting the old highway. Similar to the way Kawika Winter saw the He'eia Fishpond as a way to restore a broader Hawaiian connection to natural resources, Kāne dreamed of building a canoe as a larger metaphor for the culture. "The voyaging canoe!" he wrote in *National Geographic* in April 1976. "It lay at the very heart of Polynesian culture. Without it, there would be no Polynesia. As an artist, a sailor, and an amateur anthropologist, I had come to regard it as the finest artifact that the Polynesians had produced."[34]

Kāne, Finney, and a handful of interested others created the Polynesian Voyaging Society in 1973. As part of Hawai'i's contribution to the 1976 U.S. bicentennial celebration, the Polynesian Voyaging Society declared its intention to build a canoe and sail it to Tahiti. They decided to name the canoe *Hōkūle'a*, "star of gladness"—the Hawaiian name for Arcturus, the highest star over the "big island" of Hawai'i. Then they set to work. As Kāne predicted, it took relearning traditions far beyond navigation. "There was much to be done," he wrote. "The canoe would have to be built to ancient design. Launching it would require the revival of old ceremonies. There would be training in early Polynesian seamanship, navigation, astronomy, and craft skills. Long-forgotten food preservation techniques would have to be studied. We'd need to know much more about the old Hawaiian ways of farming, animal husbandry, and fishing, for our ancestors carried plants and animals with them and caught fish during voyages."

One year later, they had logged more than 1,000 miles sailing around the Hawaiian islands in the *Hōkūle'a*. They needed only the master navigator. At that time, however, no one in Hawai'i still had the knowledge. By one estimate, only six people left on the planet still held onto the tradition. Mau Piailug, a Micronesian navigator from the tiny island of Satawal 400 miles south of Guam, had learned the skills and was young enough to still voyage. He agreed to lead the Hawaiians to Tahiti and to train others in the old tradition.

On May 1, 1976, the *Hōkūle'a* departed from Honolua Bay, on the northwest coast of the island of Maui, and turned south to sail to Tahiti. One month later, on June 4, the ship sailed into the harbor at Papee'ete, Tahiti. More than half of the island's population crowded into the shallow water to welcome the *Hōkūle'a*. "*Hōkūle'a* demonstrated that formidable distances could be traversed relatively quickly," anthropologist Alice Beck Kehoe writes, and thus supported the oral history traditions of long-distance voyages and invasions.[35]

"When *Hōkūle'a* came it changed everything," Hawaiian navigator Nainoa Thompson said in 2022.[36] Thompson was among the thousands crammed into the Tahitian shore to see the canoe arrive. After the *Hōkūle'a* returned to Hawai'i, Mau agreed to stay to train Thompson in navigation and voyaging. In 1980, Thompson repeated the Tahitian voyage, sailing with Mau to support him. Between 1985 and 1987 Thompson sailed the *Hōkūle'a* to Aotearoa New Zealand. In 1999, Thompson navigated the *Hōkūle'a* from Hawai'i to Rapa Nui. As he grew more comfortable traversing the Pacific, another idea grew in his head. What about sailing around the entire world?

In 1984, while Thompson was working on the sailing directions from Tahiti to Aotearoa New Zealand, his childhood best friend Charles Lacy Veach started a voyage of a different kind at the Johnson Space Center in Houston, Texas. In 1985, Veach became Hawai'i's second astronaut. In the early 1990s, he made several trips into orbit on the space shuttle *Columbia*. On one of these trips, he smuggled aboard a rock adze of the type that ancient Hawaiians used to make voyaging canoes, hewn from a traditional quarry called Keanakako'i, high on the slopes of the Mauna Kea volcano. As the shuttle passed over the Big Island, Veach floated the adze past the window and snapped a picture.[37] Later, he told his friend about the photo: "Nainoa, I have a present for you," Thompson recalls Veach telling him. "The adze represents indigenous knowledge 500 years ago. The shuttle represents the intelligence of the power of technology, and navigating our future with these tools, but using great common values." Veach told Thompson that it had changed his perspective to see the Earth from space. He told him how concerned he was for the future. And he suggested that Thompson could use *Hōkūle'a* to connect people around the Pacific and around the world, to teach them to navigate. "*Hōkūle'a* needs to know the Earth and the Earth needs to know *Hōkūle'a*," he told him.

In 1995, Veach got sick with a rare form of cancer. That fall, Thompson flew to Houston to visit his friend in the hospital. Veach wanted Thompson to promise him something. "He said, 'You promise me that you'll sail *Hōkūle'a* around the world. Because you need to help the world build better, new schools. Build new schools to give the tools to our children to be able to navigate a future that we could not.'" Thompson promised. Veach died six days later. "We lost the star now," Thompson said, "but not the idea, not the seed."

It took nearly twenty years. In the interim, the *Hōkūleʻa* logged 64,000 miles of volunteer sailing time. Mau recognized five Hawaiians, including Thompson, as "Pwo" master navigators, and those masters, in turn, have trained new navigators. These new navigators include Lehua Kamalu, a mechanical engineer who had grown up attending Hawaiian immersion school in Honolulu and became the first woman to lead captain and navigate the voyage from Hawaiʻi to California, and Haunani Kane, a navigator who has brought her knowledge of Indigenous histories and observations to her position as a university professor. Thompson and the Polynesian Voyaging Society developed a sailing plan covering 42,000 miles of the ocean over three years, with stops in 337 ports in twenty countries. In 2014, Thompson sailed *Hōkūleʻa* out of Honolulu to meet the world.

The early voyages proved such a success that in 2012, the Polynesian Voyaging Society built a second canoe, named *Hikianalia* after the Hawaiian name for the star also known as Spica. In 2018, Kamalu sailed the *Hikianalia* across the Pacific to meet the climate summit. The early voyages of the *Hōkūleʻa* had been about discovery and outreach, but by 2018 the navigators' observations and the purpose of their voyages had changed. Kamalu and Thompson talked to scientists; non-governmental organizations; private industry representatives; and state, regional, and national governments, all wanting to know what was happening on the open ocean and how better to understand it. Kamalu told them that it starts with the question, Why sail in the first place? She is constantly asked why she would spend twenty-three days at sea when she could fly from Hawaiʻi to San Francisco in a little more than four hours. The answer, she said, is that "how we go about each day, how we go from one point to another, is just as important as where you end up."[38]

There's a difference between surviving and thriving, she told the climate summit, and navigating the open ocean reveals it. The voyagers' action, she said, would be to sail, to cross the seas in the way their ancestors did, and to reforge connections to the ocean as a pathway to protecting it. The people attending the summit reframed this in policy terms: how well do marine protected areas work as a tool for conservation of the high seas and as a conservation step that meets the challenges of climate change? While protecting nearshore ocean environments presents its own set of challenges, what about protection for a place of overlapping international interests in fishing, mining, and shipping traffic? Over 80 percent of the goods bought and sold around the world are transported via the ocean.[39] Collisions between large shipping vessels and marine mammals, ocean noise, and the consequences of ocean debris all raise concerns for the health of the open ocean. As Thompson spoke, he reminded the audience that the world urgently needs climate action. *Hikianalia* had journeyed across the Pacific to lend its strength to the climate movement, Thompson said. The lives of ocean people depend on action. The lesson of Polynesian navigation, he continued, is that it clarifies the questions. "You have to pay attention," Thompson said. "To stars, moon, planets, wind, clouds, oceans, waves, and life in the sea. Our navigators will make 500 choices about trim, steering, weight and balance, everything. And then they'll make two decisions in the day, at sunrise and sunset. Where are you? And where are you going? These are the vital choices of our time."[40]

Jennifer Ruesink, a marine ecologist at the University of Washington, spent much of the first two decades of her career mucking about close to shore in wetlands, seagrass meadows, oyster beds, and tide pools. She sailed a catamaran up and down the West Coast of the United States to conduct fieldwork on nearshore systems. As she explored the coasts, Ruesink increasingly dreamed of sailing across the open Pacific with her family. She, her husband, and her young daughter spent years preparing for such an adventure—learning to sail together, equipping their sailboat with tools and supplies for long voyages, and testing the idea on short family trips.

In 2018, Ruesink, her husband, and her then twelve-year-old daughter committed to the voyage and set off on a two-year trip around the Pacific. They departed from Seattle and over the course of the journey visited Hawai'i, the Galapagos, and the Marquesas Islands. Upon her return, Ruesink talked about the trip as a way to see the world and the ocean, and importantly, to work together as a team with her husband and her daughter. To cross the open ocean, they relied on each other and on a shared commitment to collaborate through joyful and frightening times.

The voyage also gave Ruesink time for close professional observation. As they hopped from island to island, she noted changes in the flora and fauna of the islands—the number of endemic species, the biodiversity of the landscape, and the interaction and connections between nature and people. She saw different strategies for conservation, including examples of stewardship that closely linked marine protection with human livelihoods. She worried about the future of the close connections she saw with nature and about communities facing immediate threats from global issues like sea level rise. She also watched, listened, and learned from the people cultivating courage in the face of such daunting challenges. "I'm interested in hope, and how people maintain hope," Ruesink said in a 2019 presentation following her return.[41]

Ruesink's voyage began with a trip from Seattle to Ecuador along the coast, taking over a year to explore and sail. She and her family then went from Ecuador to the Galapagos Islands and then west across the Pacific to the Marquesas Islands in French Polynesia—4,000 miles from the Ecuadorian coast where they departed and about 4,000 miles southwest of their starting point in Washington. Their return trip through the central Pacific brought them from the Marquesas to Hawai'i—a 2,380-mile open-ocean transit straight through what is now called the Great Pacific Garbage Patch, before one final 2,500-mile transit to return to Washington. Ruesink said that the Garbage Patch stuck with her, as it has for so many people around the world.

Early navigators crossing the Atlantic in the 1400s noticed that the seaweed collected in one place, a sea within the sea, which they labeled the Sargasso. Over time, scientists documented five such places on Earth, circular patterns around which the oceanic currents flowed, where floating particles would often be concentrated. In the past few decades, scientists realized that the floating detritus within these gyres included increasing quantities of plastics. In advance of their trip, Ruesink had read articles describing the microplastics floating in the Pacific gyre. By the time the family set out into the Pacific, Ruesink felt that the garbage patch had become a bit of a scientific curiosity, even a travel destination of sorts. Research cruises had documented the prevalence of plastic in the North Pacific, and massive philanthropic and technical efforts had begun to attempt to solve the cleanup problem. Ruesink wondered what it looked like. Recalling papers she had read with her undergraduate classes, she estimated that the Great Pacific Garbage Patch might take two weeks to sail through.

Her estimates were correct. Ruesink documented plastic floating at the sea surface every day for two weeks straight. "A new piece of plastic (usually) visible every 10–30 seconds," she told us. The majority of the large floating plastic debris that they observed and documented came from fishing gear—lines and nets. Ruesink recalls feeling that it wasn't actually that *much* plastic compared to what we see on shore at a typical cleanup of the beach or a local park. What was sobering was the scale: weeks and weeks of floating debris, over a large portion of that blue expanse of the Pacific. Ruesink described a sinking feeling as she realized how hard it would be to remove these long-distance plastic travelers and how much more work it would take to prevent the problem of plastic pollution from getting worse.

Plastic makes up the vast majority of what is termed marine debris—litter that is deposited in the ocean. Of the 300 million tons of plastic produced every year, 14 million tons end up in the ocean.[42] In albatross breeding colonies across the Pacific, the large ocean-roaming birds die with stomachs full of plastic, their bodies decomposing until all that remains is a neatly arranged pile of human junk.[43] Plastics have made it into the guts of organisms living in the deepest trenches of the ocean, and in the water, ice, and seabird colonies of the Arctic far from the source of the pollution itself.[44] Most of what ends up in the ocean arrives from land via urban runoff, sewers, or dumping. But fishing nets, lines, ropes, and abandoned vessels, and gear used in aquaculture can all be sources too.[45]

University of Toronto marine plastic pollution researcher Chelsea Rochman told us that even a few decades ago, when she was an undergraduate student at the University of California, San Diego, she couldn't find professors who believed ocean plastic was worth scientific study. She wanted to get a PhD but struggled to find an adviser. "I was getting kicked out of people's offices," she told us. "They were telling me this is not real science."

Finally, in Rochman's first year in graduate school at San Diego State University, a group of graduate students at the nearby Scripps Institution of Oceanography found funding to sail to the open-ocean area where they thought plastic might accumulate and invited Rochman to accompany them. A press release before the cruise emphasized that, while there had been a lot of media speculation about plastic in the ocean, "scientifically, very little is known about the size of the problem and threats to marine life and the gyre's biological environment."[46] At first, Rochman said, not much turned up in their nets. But a few days into the cruise, as she was having breakfast, a group of observers called to her to come quickly. From the top of the boat, they watched as a ruler floated by—and then suddenly everything turned to plastic. "There's just a confetti of plastics going by the bow," Rochman told us. "We were just standing there. It was an emotional moment." The trip to the Great Pacific Garbage Patch was one of the first where Scripps researchers communicated with the public in near real-time about what they were discovering, via blogs, tweets, photographs, and discussion boards, a practice that has now become essentially expected of large-scale oceanographic research expeditions.[47] Rochman also described the first-person view as changing everyone's perspective. Scientists who saw the scope of the problem became much more willing to make it a research priority. Conservation groups started to realize that the amount of plastic floating in the ocean was far too large to simply clean up.

Rochman has been at the forefront of plastic pollution research ever since and works with international governments on solutions. Much of what shows up in the ocean is small or even microscopic plastic. The tiny fragments form by the weathering of plastic over time. Exposed to ocean conditions and harsh ultraviolet (UV) light, plastics tend to whittle down to smaller and smaller pieces, but they don't disappear. Studies of plastic in the Pacific indicate that 94 percent of the "pieces" of floating trash are microplastics.[48]

Rochman says that it is not just the physical object that's the problem. Plastics absorb and carry various chemicals from either their manufacture or from the marine environment. "When I see plastic, I see this cocktail of contaminants," Rochman told us. "I see what will leach off, how it interacts with what's already there."

The discovery of floating plastic in the Pacific drew attention, a search for solutions, and eventually philanthropic investors. Funding from wealthy donors has allowed inventors to create machines to "clean up" a patch of the open ocean, over the concern of scientists that such projects are inefficient, ineffective, and possibly damaging to marine life. So how to clean up a problem in an ocean so vast? "One of the things that's really important to understand is that cleaning up the oceans is fundamentally different than something like cleaning up litter on the street," Memorial University geographer and plastic pollution expert Max Liboiron said in an interview in 2021. "That's mostly because of scale problems. The stuff we're really familiar with at the scale of being a human does not track into the ocean because the ocean is the biggest thing in the world. You actually have a scale problem where you cannot clean up the ocean in any way at a rate that is commensurate with the amount of plastic going into it."[49]

Even as cleanup efforts continue, scientists have predicted that the amount of plastic accumulating in the Great Pacific Garbage Patch is increasing at an exponential rate.[50] Much of the plastic may be sinking out at rates faster than we can document or collect it, meaning that the deep sea becomes the ultimate resting place for much of the debris.[51] "The best metaphor is, OK, you walk into your bathroom and your bathtub is overflowing," Liboiron said in the interview. "Do you, a) turn off the tap, or b) get a mop? I mean, eventually you'll do both, but you better turn off that tap before you start mopping up or you will never stop mopping up and you will never catch up to the water spilling out."

Much as warming and acidification have reached even the most remote areas of the globe, plastic pollution has reshaped parts of the open ocean rarely visited by people. Coastal species now take up permanent residence on floating debris, living their lives in the open ocean on rafts of plastic. Pushed by wind, waves, and current across thousands of miles, invertebrate communities can live on floating plastic for years at the surface.[52] Scientists have documented mussels, anemones, and bryozoans once considered exclusively coastal that now live in the open ocean on fishing nets and other debris.

As have so many people that we interviewed, Rochman and other colleagues in the marine plastic field have pointed out that thinking and working locally is a significant step toward addressing plastic pollution. They call for using less plastic and containing or trapping plastic before it reaches watersheds, bays, and the ocean.[53] One estimate in 2015 suggested that, of the 6,300 million metric tons of plastic ever generated, only 9 percent had been recycled.[54] Liboiron, who founded an interdisciplinary plastic pollution lab called CLEAR, has urged the scientific and conservation communities to remember the source: fossil fuel companies (which produce the petroleum for the plastic) and large multinational beverage and food manufacturers, which generate a disproportionate amount of the plastic.[55]

For many people, plastic accumulates out of sight in the open ocean. But for the people around the world who, as Polynesian Voyaging Society president Nainoa Thompson told the global climate action summit, "come from an ocean country," plastic pollution is an urgent daily threat. For the fishing communities Liboiron works with, plastic determines whether their fish is safe to eat.

Pacific Island communities contribute less than 2 percent of the plastic waste in the ocean and yet are directly and disproportionately affected by marine debris washing up on their shores.[56]

In May 2022, the *Hōkūleʻa* and *Hikianalia* sailed from Hawaiʻi to Tahiti. A joint declaration signed by Thompson and by the president of French Polynesia honored the "ancient sea road" that connects the islands across 10 million square miles of ocean.[57] Celebrations commemorated the anniversary of the *Hōkūleʻa*'s first trip to Tahiti. But the declaration and the celebrations also recognized that after forty-six years, the mission for the canoe and its voyagers had grown. The Polynesian Voyaging Society has planned a new three-year voyage around the Pacific, the *Moananuiākea*, this time to inspire "young people who can lead the many different kinds of bold voyages our Earth needs now." The joint statement with French Polynesia warned of the increasing number of threats to the ocean, from climate change to contamination. The statement concludes by committing "The People of the Canoe" to "making an urgent call for the sustainable management and protection of our oceans for future generations."[58]

To solve many of our most pressing environmental challenges like climate change and plastic pollution, we need to think of the ocean as a knowable and mappable space, not simply a void where things can happen out of sight.

CHAPTER 7

THE POLAR WORLDS

A great disconnect has grown between our communities, our economies, and our environment. This has resulted in rapid climate change that now spirals out of control and fundamentally threatens our world.
—Sheila Watt Cloutier, The Right to Be Cold

Just after midnight on December 7, 1972, the final NASA Apollo mission streaked into the sky carrying what would be the last three humans to walk on the moon. Five hours after launch, one of the astronauts looked out the window back at Earth and snapped a picture. To the astronauts, their home planet looked like a marble floating through space—and the iconic photograph, among the most widely distributed images in history, became known as the "Blue Marble." The *Apollo* spacecraft's trajectory meant it was one of the first times astronauts had a view of the south polar ice cap, and the image shows the continent of Antarctica with enough resolution to pick out surface features of the icy and mountainous terrain. Clouds swirl and thicken toward the ice-covered pole, while clear skies beyond reveal the coastlines of Africa, Arabia, and Madagascar at the center and top of the image. The "Blue Marble" shows Antarctica the way many people think of it: isolated off at the farthest edge of the Earth, blazing white and stormy in contrast to the warm earth tones of the inhabited continents.

On the surface, the Earth's poles represent some of its greatest environmental extremes. Locked in ice, fully dark several months of the year, scoured by non-survivable winds and temperatures, the Arctic and Antarctic stand for remoteness. If there is a word to sum up the poles in the Western imagination, it might be "endurance": the quality needed to survive in a harsh environment and the name of a nineteenth-century English ship (the *Endurance*) that stuck fast in the Antarctic ice, its fur-clad crew of men surviving against the elements on the strength of fortitude and resilience.

It is, of course, a vastly different story beneath the surface. The oceans all connect, and what happens in the Arctic and Southern oceans affects the others, as well as life on land far away. Cold, nutrient-rich water and long days of summer sunlight fuel incredible plankton blooms that feed fish, birds, seals, and whales whose long-distance seasonal migrations then link the polar oceans to the rest of the world. The nature of the poles is also different to people who live in the Arctic, or near the Antarctic; their stories emphasize relationships and connection rather than harshness and distance.

We have seen throughout this book the ways narratives can lead us astray. The ocean cannot tell its own story and so the story we tell about it depends on the people who get to speak. The Western ideas of the Pacific as "islands in a far sea," of the coastal Pacific Northwest as a bountiful Eden that involved little skill to maintain, and of fisheries as inexhaustible have led to environmental calamity in many cases. Now, as we have with coral reefs and kelp forests that bathe in warm water, we're finding that remoteness and treaties are not enough to preserve the polar oceans fully. People who live far from the ice see daily proof that change at the poles affects us all, including bizarre weather patterns in the lower latitudes, sea level rise, and suffering marine life. The story of the Southern and Arctic oceans now is one that the photographer of the "Blue Marble" intended to tell as they looked back at Earth from space: the story of the way nothing on our planet is truly isolated.

Of course, the story of the Arctic and Antarctic worlds starts with ice. Antarctica is a frozen continent surrounded by an ocean. The Arctic is a frozen ocean surrounded on all sides by continents. Whether it's sea ice floating on the ocean surface or land ice descending into the sea, in both places, the ice is what determines the connections to the wider world.

Scientists observe a variety of measurements of sea ice, but two are especially useful: how much area it covers and how thick it is. Both measurements, at both poles, have been shrinking for decades. In the Arctic, the extent of the sea ice traces a similar path each year: as temperatures cool in Northern Hemisphere winter, ice begins to form and build up. Sea ice extent reaches an annual maximum typically in February or March, covering 14 to 16 million square kilometers. As lengthening daylight brings warmer temperatures, the ice begins to melt around June. It is at a minimum September before the freezing begins again. In a series of popular visualizations created by climate scientist Zack Labe that he based on satellite data, the ice curves all have the same shape—but every decade the extent is a bit smaller.[1] The mean Arctic sea ice extent in the 1980s starts at 14 million square kilometers, peaks at 16 million, and bottoms out around 7 million. The 1990s starts at 13.5 million, peaks at 14.5 million, and bottoms out at 6.5 million. And so on, each curve just a bit below the next. The upside-down-rainbow symmetry of the graph belies the astonishing loss it represents. The difference between the bottom of the curve for the 1980s and the bottom for the 2010s represents the loss of 3 million square kilometers of sea ice.

The shape of the Antarctic curve is a kind of mirror opposite of the Arctic, rising through the Southern Hemisphere's winter and peaking around October before dropping precipitously and reaching its lowest extent around March or April. Like its northern counterpart, however, the overall curve keeps dropping. In March 2022, both Arctic and Antarctic graphs raised scientific alarm, with one headline in the *Washington Post* calling researchers "flabbergasted."[2] The Antarctic sea ice extent had just hit a record low in coverage, around 1 million square kilometers less than the 1980–2010 average, and parts of Antarctica were—relatively—baking in temperatures 70°F warmer than normal. Meanwhile the Arctic sea ice had recently reached its own maximum extent for the year—the tenth lowest maximum on record.

The world beyond the poles might not see the ice shrinking, but it is probably already feeling its loss. Arctic sea ice is closely related to the overall heat budget of the Earth.[3] Changing Arctic temperatures and the loss of sea ice may have contributed to the atmospheric pattern that caused extreme drought in the western United States between 2012 and 2016 and extreme cold in the eastern United States around the same time.[4] Although it's still an active and contested area of research, Arctic ice change may have ripple effects on midlatitude weather.[5] Warming in the Arctic Ocean can fundamentally reshape the way that the oceans move, circulate, and distribute heat around the globe.

In the 1960s, satellites beamed back the first photos of the full Antarctic continent. NASA lost access to some of these early pictures for several decades before regaining it in 2014—to find that there was more ice surrounding the continent in 1964 than in any picture in the subsequent fifty years.[6] It wasn't until 2011 that we got our first complete map of how the ice moves, in sheets, from the continent into the ocean.[7]

Massive ice sheets cover Antarctica. The ice flows off the land into the ocean like a giant shelf. If ice that is sitting on land melts into the ocean, it raises sea levels around the world. Tide gauges around the world show that global sea levels have risen about nine inches since 1900, although the rise has not been evenly distributed around the ocean basins.[8] As the air and water around Antarctica continue to warm, the ice sheets melt and grow increasingly unstable. Large sections of Antarctic ice shelves have already broken off, decayed, and declined.[9] Between the melting ice and the warming ocean, estimates now suggest that the United States will see as much sea level rise in the next thirty years as has been observed in the previous century, on the order of twelve inches by 2050.[10]

Shrinking sea ice causes global weather disruptions, while ice melting off the polar land leads to ripple effects around the world, including the loss of marshes, islands, ecosystems, livelihoods, and homes. In the 2018 book *Rising*, journalist Elizabeth Rush follows people around the United States who face the impossible choices caused by melting ice sheets thousands of miles away.[11] She meets and profiles lifelong coast dwellers who must choose between their history and the future rushing toward them. "I have no uncertainty about the climate science," Laura Sewall, who has lived her entire life on the coast of Maine, tells Rush. "I do have a lot of uncertainty about what to do." Others agonize and then make their choice. In Miami Beach, a homeowner talks about planning to live his entire life in his house and the realization that sea level rise meant he couldn't give his home to his kids. In Isle de Jean Charles, a rapidly disappearing island in south Louisiana where the government has tried to resettle the entire community inland, Rush talks to people who take the deal and to those who won't. Even those who agree to move describe it as a pyrrhic choice: "It's not like I threw a party when I heard about the relocation," Chris Brunet tells Rush. "I'll be leaving a place that has been home to my family for right under two hundred years." Brunet says that he thinks it's better to keep the entire community of people together somewhere else than to watch it wither slowly as the island shrinks.

In the introduction to her book, Rush visits a Rhode Island marsh where the tupelo trees have slowly started to die from saltwater inundation. They remain as standing skeletons and reminders of what's happening in far-off edges of the ocean. "The tupelos, the dead tupelos that line the edge of this disappearing marshland, are my Delphi, my portal, my proof, the stone I pick up and drop in my pocket to remember," Rush writes. "I see them and know that the erosion of species, of land, and, if we are not careful, of the very words we use to name the plants and animals that are disappearing is not a political lever or a fever dream. I see them and remember that those who live on the margins of our society are the most vulnerable, and that the story of species vanishing is repeating itself in nearly every borderland."[12]

In 2019, a year after *Rising* came out, Rush traveled to Antarctica along with scientists investigating the Thwaites Glacier. "The more I learn," Rush writes, "the more I want to stand alongside this calving glacier, want to watch freshly formed bergs drop down into the ocean like stones so that I might know in my body what my mind still struggles to grasp: Antarctica has the power to rewrite all the maps."[13]

Journalists have sometimes used the phrase "Doomsday Glacier" to describe Thwaites, the largest glacier in the world and one increasingly unstable as it warms. If all of Thwaites breaks apart and melts, its water alone could raise global sea levels by two feet. As it breaks apart it may drag other nearby glaciers with it, perhaps contributing many more feet to sea level rise.[14] Teams of scientists have been closely monitoring Thwaites like a group of physicians monitoring a declining patient. An international team of nearly 100 scientists, jointly funded by agencies in the United States and United Kingdom, monitors ice flow and temperatures, and uses robots to understand what is happening at the locations where the glacier leaves its connection to the seafloor and begins to float. For now, these scientists wrote in 2021, a floating extension of the glacier, pinned in place by an underwater mountain, helps hold the ice flow back. But the eastern edge of Thwaites is in trouble, attacked "from all angles" by warming water. Fractures have formed across the extension. Still, the scientists wrote, disaster is not inevitable, and they do not use the phrase Doomsday Glacier because "it gives the inaccurate impression that the disaster is sudden, and inevitable, and akin to nuclear war, which is not the case." Ice loss is not inevitable: it is determined by the path we choose to take.

Andrea Dutton, a geochemist at the University of Wisconsin, spent much of her career in Florida, where she says that she rarely had to convince people that sea level rise mattered—one of the most common questions she got after one of her public talks was, "Should I sell my house?" She told us that, for coastal communities, sea level rise means near-term changes in freshwater supplies as saltwater intrudes, pollution from sewage and wastewater as saltwater overwhelms infrastructure, and rising seas literally surging into streets and cities during high tides and storms. Even in Wisconsin, where she moved in 2019, she can connect sea level rise back to people's daily lives: look at the supply chain shortages caused by the COVID-19 pandemic and then imagine what will happen when coastal cities flood. Sea level rise will cripple coastal ports, airports, energy facilities, and infrastructure, the key components of the delivery of food, water, and energy in the United States and around the world. The pandemic shortages and supply chain issues are "nothing compared to what sea level rise is going to bring to our economy," Dutton told us.

Dutton has always been interested in taking a deductive approach to polar ice and sea level rise. In talks, she often introduces herself by saying, "I am a detective of the Earth." For her PhD, she studied 50- to 70-million-year-old fossil clam shells from the Antarctic Peninsula trying to figure out whether there had been ice present at the time and piecing together the climate based on what she saw. She was then offered a postdoctoral research position to study fossil corals, which record the conditions at the near-surface of the water where they lived and so can help reconstruct ancient sea levels. This research left Dutton as an Antarctic ice scientist whose research sites are based at warm, sunny beaches.

"It's a point I often make to people, that I'm studying the ice age from the tropics," she told us. "It's a nice way to think about how everything is interconnected." Dutton's fossil coral reefs are often right next to popular tourist beaches. She has one site in the Seychelles, she told us, that is the holiday backdrop for thousands of vacationing Europeans. It is one of Dutton's most valuable research sites, and it is a coral reef embedded in stable granite, like the tide pools of Monterey Bay. Dutton also works on a cluster of Caribbean sites in the Florida Keys, Jamaica, Curaçao, Barbados, Bermuda, Haiti, and the Bahamas.[15] She has traveled to West Australia, Baja California, and O'ahu. Her first step in visiting a site is often to go snorkeling so that she can get a sense of the modern reef and better understand the conditions that acted on the fossil reef.

One of the first things the global perspective reveals about sea level rise is that it is not uniform. "It's not like a bathtub," Dutton says. The age of the fossil corals combined with the geochemical records they store helps us understand the sea level at that particular island at that particular time in the past. With data from a lot of different islands, from a lot of different places, Dutton and her colleagues can start to get a more accurate picture of how high the sea level was at particular times in Earth's history. Deciphering so many difficult clues can be laborious, complex work. During a period called the Marine Isotope Stage 5e around 129,000–116,000 years ago, for example, it's clear the temperatures were warmer and global sea level was higher. In the Seychelles, the corals show that sea level peaked more than seven meters above today's levels, while at Dutton's site in West Australia it might have reached nine meters higher. But there's also considerable uncertainty. There are as many as four different explanations for how the sea behaved in that stage, ranging from a relatively smooth, stable level the entire time (just much higher than today) to dramatic oscillations with rapid peaks and retreats. The more scientists learn, Dutton told us, the more complexity they uncover.

Combining historic sea level with other clues, including ice cores from Greenland, scientists can also start to infer what was happening with the ice sheets at various times in Earth's past. In a generally warmer world, for example, seawater itself expands, which can explain a rise in sea level of several feet. A rise of more than one meter requires some melting of ice from either Greenland or Antarctica. More significant sea level rise than that essentially requires dramatic ice loss from Antarctica.[16] Scientists estimate that melting ice from Antarctica and Greenland has caused about half of the global sea level rise seen in the last fifty years, while the other half is associated with the warming and expansion of ocean waters.[17] And the melting of different ice sheets can cause rises in different places—adding to the challenge of understanding the past but also meaning that scientists have started to recognize the "fingerprint" of rises caused by individual ice sheets.

The uncertainty of the past can become a kind of transfixing mystery story. But Dutton says that, in terms of the present, we know enough. "In general when it comes to this topic, I stick to real fundamentals," she told us. In Marine Isotope Stage 5e, "sea level was higher. We're not exactly sure how high, but it was more than we could have got from Greenland. Antarctica was affected. When we talk about projections, we can argue about which curve you want to pick, why one is better than the others, but if we look at any of these, sea level is going up and it's accelerating. That's all we need to know!"

Dutton says that, at first, when she was presenting on sea level rise to audiences in Florida, she tried to avoid explaining why it was happening and just focus on what was happening and what it would mean. She soon realized that people were leaving her lectures thinking the sea was just rising normally. "It's become very clear to me that it's important to start with the why," she told us. "All my public lectures start with the why. Global warming, why it's happening, and pointing the finger at fossil fuel companies."

In her presentations, she talks about reframing the conversation, moving away from rehashing the facts and challenges and moving toward answering the question of what we are going to do. "We have an unrealistic belief that we can just engineer our way out of the situation," Dutton says. "We can build walls, and live in bowls, or we can retreat."[18] Retreat is an option made easier for people with resources and for governments willing to assist. This means that the impacts of sea level rise are disproportionately affecting those who have contributed very little to the problem of warming sea temperatures and melting ice. In *The Story of More*, geochemist Hope Jahren mentions Bangladesh, a country of around 170 million people that lies almost entirely at low levels along the delta of the Brahmaputra and Padma rivers. "If the sea continues to rise," Jahren writes, "the area of Bangladesh is likely to shrink by 20 percent over the next 30 years, crowding people into even less land and fewer resources. Incidentally, the people of Bangladesh produced far less than 1 percent of the carbon dioxide emitted to the atmosphere during the last fifty years, yet they are poised to pay the highest price incurred by its effects. This is a common trend: the people benefiting from the use of fossil fuels are not the people who suffer the most from its excess."[19]

People far beyond the frigid Southern Ocean rely on its productivity for food. Technology and industrialization now power large-scale fishing in the Antarctic, threatening the sustainability of marine resources. Booming demand for krill oil, an alternative to fish oil used in health supplements, has driven a rapid expansion of the Antarctic krill fishery. "While krill have long been used in aquaculture to fatten farmed fish, krill oil has become a gold rush for the dietary supplements industry in the past decade," journalist Richa Syal wrote in 2021 in a *Guardian* series on the changing oceans.[20] Governments have expanded marine protected areas within the Southern Ocean or used catch limits and quotas to regulate the industry.[21] As has been the story for so many fish species, however, scientists have questioned whether even "conservative" catch limits have already been too lenient.[22] Alongside the extensive history of fishing in the Southern Ocean is the parallel history of whaling—where there are krill, there are whales, of course. Whaling has occurred in the Southern Ocean since 1904 and has been more carefully regulated by international treaties since the establishment of a moratorium on commercial whaling in 1979. We are only beginning to understand the impact of the removal of the smallest portions of the global food web (krill) up to the largest (whales) and the associated impacts on carbon and nutrient cycling in the ocean.

Like the tiny copepods that determine the fate of Atlantic whales, Antarctic krill feed the biodiversity of the Southern Ocean. Michelle LaRue, an Antarctic marine scientist at Te Whare Wānanga o Waitaha (the University of Canterbury) in Aotearoa New Zealand, decided to use technology and a growing global movement of crowdsourcing scientific data to answer questions about what changing krill populations might mean elsewhere in the food web. Krill are tiny and hard to study, but Weddell seals, on the other hand, might reveal broader changes in the food web that might otherwise go unnoticed in smaller animals.

Weddell seals grow to eight to eleven feet in length and an impressive 800 to 1,300 pounds, rather large for a seal. They are circumpolar, meaning they can be found in all the areas of the Southern Ocean encircling the Antarctic continent. They are one of the best studied seal species in the world and renowned for their ability to dive to impressive depths of more than 1,900 feet.[23] The seals, which establish on-ice breeding colonies and are loyal to the colonies over their life span, live at the literal margin between land, ice, and sea on what's called fast ice, the ice that is fastened or anchored to the continent. From their icy colonies, Weddell seals head to deep waters to feed on, among other things, the Antarctic toothfish—you may know it as Chilean sea bass—which are themselves voracious predators of smaller fish, shrimp, and squid. Here is the Antarctic food web and its ripples: the seals and people eat the fish, the fish eat the shrimp and squid, which eat the krill. And the people also harvest enormous amounts of the krill.

Weddell seals have been studied extensively and are well tracked in the Ross Sea, partly because of McMurdo Station, a research station operated by the U.S. Antarctic Program and host to the largest scientific community in Antarctica, with around 1,200 residents. Scientists on the ground have meticulously tagged, counted, and observed the seals in McMurdo Sound for decades.[24] In an online seminar in 2022, LaRue said that she began to wonder if there was a way to study the seal populations that stretched all the way around the continent.[25] She had seen a high-resolution satellite image of Antarctica and noticed that she could identify what she thought were seals on the ice. Their large size and dark fur help Weddell seals in particular stand out, dark specks against a brilliant white backdrop. LaRue contacted a suite of researchers who had done long-term surveys on seals to ask: Do you agree that we can see seals from satellite photos? And, if so, could we estimate seal populations across the inaccessible Antarctic continent using satellite data? They thought it was worth a try.

LaRue and colleagues started by looking at satellite images that had been taken at the same time as the ground-based surveys done by scientists. They found a very high correlation between counts done via satellite photo and via survey.[26] LaRue was intrigued. There are more than 100,000 square miles of fast ice where seals could be seen in satellite photos. She started trying to work within her research lab but quickly realized she couldn't recruit enough students to sort through and identify seals in the many hundreds of high-resolution satellite photos of Antarctic ice. A colleague suggested turning to the power of the internet for help.

As LaRue considered her seal-counting question, other scientists were using similar approaches to investigate kelp forests and penguin colonies remotely using the powerful combination of enthusiastic online science fans and satellite imagery.[27] LaRue decided to try a system termed search area reduction: show people an image and ask if there's a seal in it. The answers, multiplied by hundreds of thousands of internet users, rapidly expanded the data set of seal locations: both where seals are and where they are not in Antarctica.

LaRue focused first on the Ross Sea, a deep bay carved out of the Antarctic continent south of Aotearoa New Zealand. Comparing the crowdsourced seal data to the data collected over years of research on Weddell seals in the Ross Sea, LaRue and her colleagues gained insights into aspects of the life of the Weddell seal that had been mysterious until then. First among them, LaRue wrote in a research paper in 2016, was how little of the fast ice the seals were actually using.[28] Seals were present on just 0.55 percent of fast ice in the images from 2011, and they were not uniformly distributed.

In general, LaRue says, online community scientists tend to overidentify seals, but they don't tend to miss seals—if seals are there, people spot them. But they also spot other things—crabeater seals, emperor penguins, rocks—and call them seals. To refine the data, LaRue decided to launch a second campaign that asked participants to circle individual seals on the images. An algorithm tracked agreement on whether an individual speck on the ice is a seal. The more people who agree, the more likely that the speck is a seal.

In a summary paper in 2021, LaRue and colleagues made the first-ever estimate of the global Weddell seal population: 202,000. Even more, the quality of the crowdsourced data combined with habitat modeling allowed them to begin to explain why the seals live where they live and what makes that small percentage of fast ice so important.[29] The data offered insight into how far Weddell seals like to be from shore while they are on the ice (a little bit inward on the ice, possibly to be away from killer whale predators) and the preferred distance from Adelie penguin colonies (far!) and Emperor penguin colonies (near, unless the colony is too big). The spacing of Emperor penguins across the ice and the corresponding spacing of Weddell seal colonies may be driven by competition for food.

The next step, LaRue said, will be to work backward through satellite images from the past to track seals through time. As with every habitat in the ocean, it is hard to see change without a baseline, and the matter of Antarctic baselines has been more challenging than most. The more researchers learn about the seals, the more they can make conclusions about what is happening across some of the least accessible research terrain on the planet. In papers, LaRue often compares the crowdsourced Weddell seal effort to a previous effort to estimate the population of Antarctic crabeater seals. The Antarctic Pack Ice Seal Program involved researchers from a half-dozen countries using ship and aerial surveys over several years. The crowdsourced "seal or not a seal" project searched Antarctica at three times the rate and at a very, very small fraction of the cost. In doing so, it helped thousands of people online gain a better understanding of the scientific method and the marine ecology of a remote continent.

A new, efficient, holistic method for counting seals comes just in time, LaRue and her colleagues wrote in 2021. The Weddell seals are indicators of the health of the broader Antarctic ecosystem. The seal project "confirmed a continent-wide, interactive relationship" between Weddell seals and other animals. Because they prey on a commercially fished species, the seals can indirectly reveal fishing pressures. Because they rely on a small area of fast ice, they are sensitive to environmental change. And the Southern Ocean is already feeling the effects of climate change.[30] While sea-surface temperatures are increasing, the extent of water covered in sea ice has shrunk, and the depth of the ice has decreased. These and many other changes will likely alter the pattern of phytoplankton blooms and thus the entire Antarctic food web. "If predictions of change in the Southern Ocean over the next 100 years correctly forecast warming sea surface and air temperatures, altered winds and sea surface salinity, ice shelf loss, and increases in invasive species among others," LaRue and her colleagues conclude, "then large-scale approaches such as ours to monitor the Antarctic ecosystem will be of increasing importance."[31]

In a close parallel with the krill-rich Southern Ocean, changes in the amount of sea ice cover directly affect phytoplankton and zooplankton in the Arctic Ocean, leading to changes throughout the entire food web. The Chukchi Sea is an extension of the southern Arctic Ocean, in the space formed between continents where the northwestern slope of Alaska and the northeastern edge of Russia tilt toward each other. Below it, the Bering Strait connects the Arctic to the Pacific Ocean. Placed at the great crossing ground between oceans, icebound in winter and flush with food in the spring and summer, the Chukchi Sea is among the most productive regions in any ocean and home to enormous migrations of marine mammals.[32]

Thinner sea ice causes a cascade of effects. White ice reflects the sun's rays, so less ice means a darker ocean, and a darker surface for solar radiation to be absorbed, further heating the Arctic. Sea ice can also act like a lid on the Arctic Ocean. When the lid comes off, storms and winds stir up the water, mixing layers and increasing the heat exchange between the ocean and the atmosphere. Mixing brings nutrients to the surface and, combined with an ice-free surface layer that allows more light to penetrate the ocean, can drive increased phytoplankton blooms, with sometimes novel species. Soon, the zooplankton, the next step in the food web, respond and the effects reverberate.

The Chukchi Sea and the Arctic Ocean in general have what scientists call a short food web, meaning there are only a few steps between the phytoplankton at the base of the food web and the top predators. Arctic marine mammals—polar bears, beluga whales, narwhals, bowhead whales, ringed seals, bearded seals, and walruses—will likely feel the changes quickly. (Of course, Arctic human communities also rely on this food web and will also feel the impacts of each disturbance.) Researchers tend to split Arctic mammals into two groups: those who are ice obligate, which means they live on the ice for some part of their lives, and the ice provides a platform for foraging, reproduction, caring for young, and resting, and those who are ice associated, meaning the presence of the ice influences their foraging, migration, and behavior.

Donna Hauser, a marine ecologist at the University of Alaska Fairbanks, studies multiple genetically distinct populations of ice-associated beluga whales near the Chukchi Sea. The whales spend winter in the North Pacific, then pass through the Bering Strait in the spring to spend the summer in the Chukchi or Beaufort sea. Hauser has worked for years with scientific colleagues and local communities to tag and track beluga whales carefully. Every time the whale comes to the surface for air, the tag transmits data to a satellite and then back to the scientific team. In a presentation in 2019, Hauser showed animations of dots on an Arctic map, each purple or green dot represented a whale swimming around the Chukchi Sea. Tags are usually attached in July; if the scientists are lucky, they might get more than a year of data from an individual whale. More than sixty whales have been tagged and watched over decades of research.[33]

Hauser wants to know whether the changing extent of sea ice determines where the whales travel and forage for food. One lesson from watching the data, however, is that the whales like to move around. "Even at a very coarse scale . . . you can see that it's a quite mobile habitat," Hauser says. "It's quite an interesting seascape for these animals." Using the tagging data and theoretical models, Hauser tries to understand the ideal habitat for the whales. Sea ice was not one of the primary predictors of whale location, but there were other factors, including oceanographic conditions and bathymetry that were strong predictors. Hauser and her colleagues do not think that the changing locations of sea ice are yet influencing where the whales tend to go. So far, the tags show the whales tend to return to their same favorite spots, year after year, looking for prey.

When she compared the 1990s to the 2000s, however, Hauser found that the whales seem to be diving longer and deeper in the summer. The shift might be an indirect effect of the sea ice changing and where their Arctic cod prey is going in response. Hauser was also interested in understanding whether the seasonality of freeze-up in the Arctic might influence whale migration patterns. So she and her colleagues investigated the timing of the beluga whales passing specific points in the Arctic during their fall migration. They documented a three-week difference in the median date of migration for Chukchi Sea beluga whales. It appears that the shifting date of freeze-up is opening habitats and modifying migration patterns. A surprising finding was that the whale population just next door, the Beaufort Sea belugas, showed no significant change in the time when they passed the key migration points, indicating a strong fidelity to their centuries-old patterns. It is an interesting time to be a beluga whale in the Arctic.

In the presentation about whale migration patterns, Hauser showed an image of Roald Amundsen, an early twentieth-century Norwegian polar explorer. The black-and-white headshot of Amundsen wearing a fur-lined coat and looking directly into the camera is superimposed on a photo of the *Gjøa*, his sixty-foot wooden sailing ship. Between 1903 and 1906, Amundsen and his crew navigated the Northwest Passage, the sea route that connects the Atlantic and Pacific oceans along the often ice-blocked northern coastline of North America. Modern paleoclimate reconstructions suggest that the summer sea ice in the Arctic Ocean at the time of Amundsen's voyage covered 3.8 million square miles.[34]

By the 2020s, the summer sea ice extent in the Arctic hovered around 1.5 to 2 million square miles, the lowest it has been in both the measured instrumental record and the reconstructed paleoclimate record going back thousands of years.[35] Scientists predict that the Arctic could see summertime ice-free conditions—a sea ice extent of less than 1 million square kilometers (roughly 386,000 square miles)—by the 2030s or soon thereafter.[36] Longer paleoclimate records that tell the story of Earth's climate evolution and history indicate that these seasonal ice-free conditions have not existed in millions of years.[37] In 2016, an 820-foot cruise ship carrying about 1,000 passengers completed the Northwest Passage route in three weeks. One year later, a tanker traversed the Russian Arctic coast without the use of an icebreaker escort.[38] Taking the northern routes through the open Arctic might save weeks in travel times compared to routes through the Suez Canal, so declining summer sea ice will reshape global shipping.

Like North Atlantic right whales and eastern Pacific gray whales, Arctic marine mammals now face dramatic encounters with vessel traffic even as they shift their own migrations and follow food into new places. The preliminary comparisons are alarming, Hauser notes. Studies from the North Atlantic show that, even in addition to the risk of ship strikes, the noise from vessel traffic interrupts whales' use of sound for finding food, communicating, and locating a mate. The Arctic bowhead whale is one of the few whale species that lives entirely in the Arctic and sub-Arctic, and among its closest relatives is the North Atlantic right whale. Hauser hopes learning more can prevent the bowhead whale from following the right whale's declining numbers and so has tried to model the overlapping factors of whales' exposure to vessel traffic and their sensitivity to the presence of vessels.[39]

A warmer world means more parallels between the Arctic and its cousin oceans to the south, including in the water itself. The Arctic's thick cover of sea ice creates a cold layer of fresh water, called the halocline, just beneath the surface of the ice. Because fresh water is less dense than salty ocean water, the halocline floats over the top of the warmer, saltier Atlantic currents that push northward, insulating the ice above it. Studies show that the halocline is weakening, however, and the Atlantic water pushing in has more and more heat.[40] The warmer water mixes toward the surface, making the Arctic Ocean more like the Atlantic itself—a process scientists have labeled the Atlantification of the Arctic.[41] On the other side of the Arctic, scientists have seen the flow of water through the Bering Strait and into the Chukchi Sea nearly double since the 1990s—leading to Pacification as the Pacific's warmer, relatively fresher waters flood north.[42]

Scientists also refer to the Arctic becoming more like the Atlantic and Pacific as borealization and have been surprised at how fast and how deep the invading ocean currents have arrived as sea ice declines. As with marine species at all latitudes, the warming pushes traditional Arctic species toward the poles—but this far north already, the potential habitat to escape into grows smaller and smaller. Meanwhile borealization represents a great new opportunity for species that have long found the Arctic cold limiting. Consistent with what Malin Pinsky found on the eastern seaboard of the United States, fish follow the reorganizing ocean currents into new places in a way that matches the rate of change in the water itself. In the Barents Sea, the vast shallow sea north of Europe, the entire food chain is becoming more like the ecosystems found outside the Arctic.[43] The same is happening around Alaska and Canada as Pacific water surges into the Chukchi Sea and causes widespread ecosystem reorganizations.[44]

In the fall of 2019, a shift, a collapse, a regime change—scientists still don't know what to call it—rippled through the seabird populations in the Chukchi Sea. Seabirds—especially short-tailed shearwaters, a dusky gray bird abundant around the world, as well as puffins, murres, and auklets—began washing up dead by the thousands on Alaskan beaches. Although seabird die-offs have happened before in the Chukchi Sea region, the number has increased, with large die-offs every year since 2015. Researchers collected seabird carcasses and found that many had starved to death.[45] Warm-water conditions for several years had affected the small fish that the entire Arctic—from people to whales, to birds—rely on for food. By 2018, National Oceanographic and Atmospheric Administration (NOAA) monitors saw an increase in mortality and strandings of bearded, ringed, ice, and spotted seals. The cause of this marine mammal die-off is still undetermined—although it's likely a combination of ecological factors that have led to starvation and potential exposure to harmful algal blooms.[46] During the same time, a group of scientists collaborating on research in the Bering Strait and Chukchi Sea documented the first observation of a harmful algal bloom that becomes concentrated in shellfish, from an algae once thought to be limited to more southern waters.[47] Scientists have tracked the same toxin moving from shellfish into marine mammal populations.[48] After several years, the causes of the die offs, warm water conditions, and harmful algal blooms remain uncertain. Researchers have been slow to investigate the seabird and marine mammal die-offs relative to the urgency felt by people living around the Bering Sea, who rely on a healthy Arctic ecosystem. "It's not an academic issue out here," Sea Grant biologist Gay Sheffield told the *Seattle Times* in 2019. "If you live here, it's a real-world problem. It's an immediate, stop, get an answer problem. It's a food security and health problem."[49]

As scientists try to understand the rapid changes in the Arctic, some researchers and funding agencies have begun using the name "the New Arctic."[50] In 2018, the U.S. National Science Foundation (NSF) announced it would fund Arctic research as one of ten "Big Ideas," with each receiving $30 million.[51] In a 2018 letter announcing its Navigating the New Arctic grants, the NSF "particularly encouraged" proposals that involved community engagement and coproduction of knowledge with Indigenous communities.[52] Because most Arctic research is conducted by people from outside the Arctic, scientists from across the United States reached out to people who could facilitate the cross-disciplinary, cross-cultural communication that the grants required. Many of those projects found Kaare Sikuaq Erickson. Erickson is the son of a Norwegian father from the Norton Sound village of Unalakleet and a mixed-race Iñupiat mother from the North Slope community of Utqiaġvik, near the northernmost point in the United States. Early in Erickson's life, the family moved often. They lived for a while in a Siberian Yupik village on St. Lawrence Island, a remnant chunk of the Bering Sea land bridge that is closer to Russia than it is to Alaska. They lived in St. Michael, a 400-person village on the west coast of Alaska that is the northern edge of Central Yupik territory. They lived mainly in what Erickson considers his hometown, Unalakleet, now considered the southernmost Iñupiat village. They visited family members who lived farther north, in Shishmaref and in Utqiaġvik.

Erickson went to college in Anchorage and received undergraduate and master's degrees in anthropology. He learned along the way how to navigate between Western and Indigenous traditions. Beginning in 2003, he worked on a series of projects around Alaska, helping researchers visit village sites, translating and doing transcriptions, planning Arctic logistics, eventually joining the Ukpeaġvik Iñupiat Corporation as a scientific liaison. When the NSF made engagement with Indigenous communities a requirement, Erickson had the background that researchers needed. Erickson advised NSF-funded scientists on research proposals; connected them to Indigenous communities; and helped identify, describe, and communicate the changes happening across the Alaskan Arctic. In 2020, he published his master's thesis, a close look at what he characterized as examples of successful engagement between Iñupiat and academic scientists in Utqiaġvik, and what made it work.[53] He also established his own company to help build cross-cultural connections throughout Alaska, which he named Ikaaġun, after the Iñupiat word for "bridge." "I use my upbringing in everything I do," Erickson told us.

Erickson takes the long historical perspective on how Arctic people have survived, thrived, and adapted their cultures, tools, and ways of life to change. In a recent presentation to the Arctic Research Consortium, he showed a series of maps documenting archeological evidence for the human migration into and occupation of what is now known as Alaska, emphasizing the evidence of human settlements along the coastal route.[54] "We all know this 'land bridge' theory," Erickson says. "A lot of us were taught this in schools growing up . . . but now we are starting to realize that people were smarter than we thought. They were traveling in boats much earlier than that."

By their very definition and location, coastal settlements have been susceptible to erosion and even complete submergence under rising sea levels. For archaeologists who study human migration, it means the archeological record ended up biased toward inland sites. For Indigenous Alaskans like Erickson, it also shows that shoreline change has been fundamental to living in the Arctic for as long as people have lived there. In his presentation, Erickson described Alaska and the Arctic as a crossroads of continents, of ocean and land, of trade routes both historical and modern. Research proposals may focus on the New Arctic, and change might be affecting everyone, but Erickson emphasizes often that the changes, risks, and adaptations vary enormously. He sees rapid change in the context of millennia of Indigenous history in the Arctic, sometimes objecting to the framing of the research itself. "People assume that we're entering this new Arctic, when in reality we have faced adversity for thousands of years," Erickson told a writer from *High Country News* in 2021. "We've always been able to adapt and be resilient. This is no different."[55] Some villages face existential threats without the resources to meet them, yet other places have developed extensively and might be able to weather some new extremes. One village could be on a barrier island, one could be on a solid foundation, one on a riverbank. Rising sea levels, coastal erosion, and extreme weather will affect each differently. Even the climate itself within an area as vast as the Arctic is so variable that changes do not mean the same thing everywhere.

Kivalina is a village of around 450 people that was built on a barrier island in the Chukchi Sea. It is eighty miles north of the Arctic Circle on the northwest coast of Alaska—a stretch of coastline that is across the sea to Siberia. In a story that is repeated across the Arctic, the village site was traditionally a temporary stopping point for Indigenous Iñupiat hunters, but it was turned into a permanent settlement when the federal government built a school there. Today it has a rectangular grid of streets and houses, plus a church, school, and post office. Changing sea ice has brought the village to a crisis.[56] A later freeze-up means less ice in the fall months, when the Arctic's strongest storms arrive. Without the protection of the ice, the sandy island's shoreline has been battered, causing extensive erosion and damage to Kivalina's infrastructure. Sea level rise has made the problem worse. In 1998 and again in 2000, residents voted to move the entire village.[57] But federal disaster relief doesn't cover relocation or even damage caused by erosion.[58] Without financial support, Kivalina's residents have been unable to move.[59]

Similar stories play out in the Canadian Arctic, where communities are making plans and requesting funds to move away from growing coastal hazards from climate change. The Canadian Arctic coastline includes some of the fastest eroding places in the world, with boulder-size chunks of coastal cliff detaching and falling into the ocean as the permafrost warms and ice retreats. Permafrost, ground that remains frozen for multiple years at a time, accounts for nearly one-quarter of the land surface of the Northern Hemisphere and is a massive storage of biomass carbon.[60] Charlie Paull, a scientist at the Monterey Bay Aquarium Research Institute, told us in 2020 that he had recently made a two-week trip to the Canadian Arctic to document shoreline changes. Across his research sites, he saw retreating ice, slumping shorelines, and permafrost melting into intertidal habitats. The Arctic is still moving out of its last glaciation, so some of those environmental changes might be expected to take place slowly over long time periods. But some of it is also rapid, human-caused change superimposed on the long-term pattern, and Paull wanted to tease apart the two. Like so many other ocean researchers in so many places, he was worried about the baseline shifting before research could catch up. "It is going to change rapidly," he said. "Before we even know what is going on."[61]

Permafrost often extends from the land out into the ocean, where it has been inundated as sea levels rise. Two years after we spoke to him, Paull and his colleagues published a paper showing dramatic changes to the Arctic seafloor as the submarine permafrost thaws. Using a variety of tools to map the subsurface, including autonomous underwater robots, they showed deep sinkholes and pockmarks forming in the sediment over a nine-year period. The paper calls the changes "extraordinarily rapid"—and, echoing what Paull told us, noted that it is hard to know how widespread such changes might be on the Arctic shelf because this was one of the first times researchers had been able to map it.[62]

Kaare Erickson told us that he had recently been asked to write a short article for an academic journal about the changing Arctic through the eyes of an Indigenous person. At first, it seemed an impossible request: summarize changes across three seas, separated by thousands of miles and spanning ten degrees of latitude, and home to hundreds of diverse villages. "We are all dealing with similar issues associated with a changing climate," Erickson says. "However, we are all forced to deal with them differently because of the varying socioeconomic issues across the Arctic."

He decided to focus on the experience of change in one place, his hometown Unalakleet, where the loss of sea ice, warmer and wetter weather, shifting animal migrations, and permafrost thaw have interrupted daily life. "The people of Unalakleet, like those in every community in the Arctic, will continue to face and overcome challenges as they arise," Erickson wrote. "The question is: are any of these challenges avoidable?"[63]

When we talked to him again in 2023, Erickson had just started working on a new four-year project, funded by NSF, that will create what the organizers call a National Indigenous and Earth Sciences Convergence Hub, with research taking place in partnership with Indigenous communities in Alaska, Louisiana, Puerto Rico, and Hawai'i.[64] Erickson was also trying to create a new organization based in Anchorage to lead immersive Alaska orientations for scientists, teachers, and business teams who want to work in partnership with people of Alaska. He aims to create multimedia content to help explain the rich history of Alaska's diverse Indigenous communities, and lead tours—berry picking, salmon fishing, kayaking.

On an interconnected planet the change does not just flow in one direction outward from the ice. What people do elsewhere can lead to sometimes alarming change at the poles. In 2017, Chelsea Rochman and her colleagues surveyed the Arctic waters of Hudson Bay and the Nunavut archipelago, collecting surface water, snow, zooplankton, and sediments from the northernmost stretch of Canada. More than 80 percent of the samples contained microplastics.[65] Rochman's group compared the samples they took with the population density on land nearby and found no connection—leading them to conclude that the plastics are being transported over long distances to the Arctic. Her discovery of microplastics blowing into the Arctic and accumulating there mirrors a discovery that the heavy metals from industrial activities around the world accumulate in the otherwise pristine Arctic, with dire impacts on Canadian Arctic Indigenous communities.[66] The deposition of arsenic, mercury, and lead are well documented in the Arctic, with long-range atmospheric transport as the key mechanism.[67] The presence of these heavy metals and their accumulation in the marine food web, including subsistence food sources, is a continued assault on the Arctic and its people. And now, thanks to Rochman's research, we know that plastic fibers are making it to these remote locations as well. And not just plastic: Rochman's research documented cotton microfibers too, dyed blue, probably derived from jeans or similar clothing items. Even if we never get to set foot in the Arctic, our footprints are there. A decrease in sea ice cover is expected to increase plastic pollution in the Arctic, first through the deposition of microplastics that are currently tied up in ice but ultimately through opening the Arctic to increased tourism, shipping, and human impacts overall. There are even concerns that floating plastics will bring invasive species to an already rapidly changing ecosystem.[68]

Some southern Māori oral traditions tell the story of a great Polynesian navigator, Hui Te Rangiora, who sailed his fleet of canoes south from Rarotonga into a frozen ocean. In 1899, a white ethnographer named Percy Smith, who worked among the Māori and edited the *Journal of the Polynesian Society*, published the story.[69] He described another celebrated Polynesian navigator who wanted to follow in Hui Te Rangiora's footsteps to see "the rocks that grow out of the sea, in the space beyond Rapa; the monstrous seas; the female that dwells in those mountainous waves, whose tresses wave about in the waters and on the surface of the sea; and the frozen sea of *pia*, with the deceitful animal of that sea who dives to great depths—a foggy, misty, and dark place not seen by the sun. Other things are like rocks, whose summits pierce the skies, they are completely bare and without any vegetation on them."

Smith wrote that there could be no doubt what this quotation meant: Hui Te Rangiora had visited Antarctica. The tresses were bull kelp; the deceitful animal a seal or walrus; and *pia* meant arrowroot, which, Smith argued, was a clear way of describing sea ice for Polynesian navigators who'd never encountered it before. In a 2021 article, New Zealand biologist Priscilla Wehi and several Māori colleagues used Smith's translation to argue that Hui Te Rangiora was the first person to see Antarctica, 1,000 years before any Westerner. And, they continue, the Māori have been present, although unacknowledged in Western accounts, for the entirety of human history in Antarctica.[70] Archaeological evidence suggests that shortly after arriving in Aotearoa, the Māori explored and settled in islands throughout the Southern Ocean. By the fourteenth century they had visited Maungahuka, that is, the Auckland Islands, nearly 300 miles south of the southern edge of New Zealand and one-third of the way to Antarctica. Their motivation, Māori historian Michael Stevens wrote in 2021, was likely scientific curiosity about the world around them: "The speed of dispersal from mainland New Zealand and the consistency in species targeted for consumption throughout South Polynesia strongly suggests that the settlement of outlying islands was not driven by population density or competition for resources, so-called push factors, but by a desire to fully explore and take stock of the region."[71]

In the 1800s and 1900s, the Māori joined European explorations of the Antarctic continent in a variety of roles and continued through the European age of exploration, with navigator Te Atu the first Māori to view the coast of Antarctica in 1840. Māori men held valued roles in fishing and whaling voyages in the Ross Sea, Wehi writes, "despite a backdrop of discrimination and racism." After the human presence on the continent shifted toward science in the mid-twentieth century, scientists of Māori descent have lived on the continent's research stations and conducted research in physics, biology, and Earth science. Scott Base, New Zealand's scientific research station on Antarctica, is decorated with carvings and artwork that celebrate exploration; navigation; and mātauranga, or traditional Māori knowledge. "Over the last 200 years, Antarctic narratives have contributed to conceptions of Imperial adventure, carried out by predominantly European male explorers," Wehi writes. "Increasingly, however, researchers and communities are re-examining the underlying paradigms that shape this thinking, including issues of racism and colonialism, and asking what an inclusive Antarctic future might look like."[72]

In 2020, a group of southern Māori began a project to document Māori connections to the Southern Ocean and Antarctica and to call for a new direction in New Zealand's Antarctic policy. Through a series of research reports, surveys, community meetings, and media, the researchers tried to emphasize connections between Antarctica, the Māori, and the world's oceans. In a 2020 community webinar, Sandy Morrison, the acting dean of the University of Waikato's Faculty of Māori and Indigenous Studies, asked viewers to flip the world atlas 180 degrees. Instead of viewing a chopped-off Southern Ocean and Antarctica at the very bottom edge of the map, she said, we should think of it as the top of the world. From there, she continued, is a clear path to make connections to the continent and to navigate a future in which Māori concepts inform climate change adaptation.[73] It is a historical coincidence that in the iconic "Blue Marble" image of Earth from space, the original photograph, from the astronaut's perspective, had Antarctica at the top. The version released by NASA and spread around the world—the definitive image of Earth's fragility and isolation—was cropped and rotated to match the north-is-up perspective of people in the Northern Hemisphere.[74]

"We have so many stories of our people chasing whales, following octopuses, working with the dolphins, working in seasons with certain species, following the albatross," Morrison said. "We have stories that tell us our ancestors also traversed into Antarctica and the Southern Ocean. What name they gave to it we know not. But we do know through the stories that came back to us that they were there. In this association with Antarctica we are cementing ancient relationships between us as Māori and also Pacific people, but also with place—and place includes Antarctica."

For 106 years after it sank in the ice-choked Weddell Sea, the English ship *Endurance* sat at the bottom of the Antarctic seafloor, 10,000 feet below the surface in the ice-cold darkness. The ship's captain had left detailed records of where the ship became engulfed by ice and where it sank, and for decades archaeologists had hoped to find the wreck. As deep-sea technology improved, the discovery became more and more possible. Finally, in March 2022, after years of planning and fundraising; thirty-five days at sea; and two full weeks driving submersibles, drones, and other equipment around a 150-square-mile area in treacherous conditions, a team of English archaeologists announced they had found the *Endurance*.

The first images to appear online showed a ghostly ship emerging from clear Antarctic waters: details of the metal and wood railings, the intact rear deck wheel, and a nearly pristine-looking stern with the clear lettering of ENDURANCE across the top. The ship appears exceptionally well preserved; the very cold waters of the Southern Ocean do not harbor wood-eating microbes or invertebrates.

An entire ecosystem of deep-sea species has made a home on the wreck. In footage released by the Endurance22 project, Antarctic sea anemones crowd the space above the ENDURANCE lettering, tentacles outstretched. Sea stars with dangling arms catch plankton in the currents, next to glass sponges, and stalked crinoids inhabit the deck.[75] People have previously compared deep Antarctic ecosystems to a glimpse into ancient oceans because they are dominated by invertebrates, with only a few fish species that are well adapted to the icy conditions.[76] Seafloor biodiversity in the Southern Ocean is partially determined by the presence of boulders or rocks in the sediment that provide hard locations for invertebrates to attach to. Shipwrecks make a good alternative.

If the ship remained well preserved, the conditions above it and the attitudes of the people who found it had changed dramatically. Mensun Bound, the director of exploration for the team that found the wreck, wrote in 2022 that he hoped inspiration could be drawn from the wreck to help confront a polar world in flux. "It is not all about the past," Bound wrote. "We are bringing the story of Shackleton and *Endurance* to new audiences, and to the next generation, who will be entrusted with the essential safeguarding of our polar regions and our planet. We hope our discovery will engage young people and inspire them with the pioneering spirit, courage and fortitude of those who sailed *Endurance* to Antarctica."[77]

CHAPTER 8

THE DEEP

The intertidal turns over in a decade. The deep sea turns over in a century.
—Jim Barry, interview with authors, November 14, 2017, Monterey, CA

When you stop at the rocky edge of the Hewatt transect and the tide pools of Pacific Grove, California, on a clear day, you can look across the blue sheet of Monterey Bay to the purple ridgeline of the Santa Cruz Mountains. From point to point, it's about twenty-five miles, encompassing a natural bay that opens on the Pacific and closes on the California shoreline like a curly bracket. The region's long-running relationship with the ocean, from the riches of its tide pools to the once-massive sardine fishery, to the popular whale-watching tours that still leave the harbor today, is a consequence of Monterey Bay's utterly remarkable underwater geography. About ten miles north of the tide pools, in the middle of the bay, the seafloor bottom suddenly drops into the abyss. This point is the entrance to a long, twisting canyon, its walls sheer rock that reach a full mile high in places. The canyon snakes through the center of the bay from nearly the edge of the seaside dunes to 290 miles offshore. The Monterey Submarine Canyon compares in shape and depth to the Grand Canyon. Constantly scoured by underwater avalanches, the canyon fans out into the ocean's abyssal plain at a depth of more than two miles beneath the surface.

When Jim Barry wanted to take what he had learned from tide pool transects to explore the deep ocean, he did not have to go far. The Monterey Bay Aquarium Research Institute (MBARI), where he has worked since 1991, started running underwater expeditions to the canyon in the 1990s with its new remotely operated vehicle (ROV) *Ventana*, which is launched mid-bay from a research ship. The cruises left from Moss Landing, inside Monterey Bay, and dropped into the canyon in the protected waters of the Monterey Bay National Marine Sanctuary.

First developed for military and then industrial purposes, ROVs have powered a revolution in deep-sea exploration over the last half-century. They remove the human element inside the submarine and place the scientists and pilots safely in the control room of a vessel at the surface. A crane lifts the ROV off the boat and onto the water's surface, where it begins to descend to its destination. A large scientific crew can observe everything the ROV sees on its cameras via television screens around the ship. These images can also be sent via satellite in real time back to shore, enabling a shore-based collection of scientists to serve as observers. ROVs are tethered to the ship, but scientists can also use autonomous underwater vehicles (AUVs), which travel underwater using preprogrammed routes, without the restrictions of a tether or a pilot at the surface.

The deep sea is often characterized as the last frontier on Earth. Only a small fraction of the ocean floor has been mapped at high resolution (about 20 percent, depending upon how you define "high resolution"), although the rate of exploration is accelerating. Some scientific estimates suggest that as many as 90 percent of the ocean's species remain to be identified, many of them because they have hidden in the deep ocean where humans still rarely linger.[1] As has been the case from tide pools to fishing grounds, to the open ocean, however, the frontier is the site of a race between explorers who chart and document and those who are ready to exploit what's newly mapped—or even yet to be mapped. We understand, relatively speaking, little about this habitat on Earth, and what we have learned thus far has been through fairly recent technological advances. What we do know for certain is that the deep sea is fragile, slow to recover from disturbance and filled with discoveries yet to be made, and surprisingly connected to and influenced by human activities at the surface.

On a routine ROV dive in June 2004, Barry was watching the screen when something unusual emerged from the darkness.[2] The *Ventana* was exploring an area three-quarters of a mile below the surface, looking for a sediment trap instrument. The seafloor was flat and muddy, bristling with weird worms and devoid of structure. A sharp edge appeared first on the video. It was an obvious industrial shape, metallic and boxy. As the rover hovered, the shape resolved into a shipping container. It was lying upside down and at a slight tilt, its corrugated underside collecting shells and the gently falling marine snow. The container had fallen in a position where Barry could still read its serial number, which he dutifully wrote down. Next to the container's handle, shining in the beam of the *Ventana's* lights, was a clearly legible bright yellow caution sign.³

Barry and colleagues turned over the serial number on the shipping container to the Monterey Bay National Marine Sanctuary, which in turn reported it to U.S. customs. The agencies tracked it back to a ship called the *Med Tapei*, a 40,000-ton cargo ship built and registered in the Netherlands and operated over time by three multinational shipping companies. An investigation revealed that on February 24, 2004, the heavily loaded *Med Tapei* ran through a winter storm near the California coast on its way from San Francisco to Los Angeles. Just before 1 a.m., fifteen containers broke loose and fell overboard. Federal investigators later found that the containers had been improperly loaded, with bad welds and missing anchor rings.⁴ The fifteen containers, holding tires, plastic items, cyclone fencing, hospital supplies, wheelchairs, clothes, and recycled cardboard, sank into the Monterey Bay National Marine Sanctuary. It was only three months later when Barry found one of them. The serial number on the found container indicates that it is full of Michelin tires. It was buried too deep to consider retrieving it, so instead the shipping company's operators paid a fine to the National Oceanographic and Atmospheric Administration (NOAA). The other fourteen containers have not yet been found.⁵

It would be nice if stories like Barry's were isolated incidents, but they are common in deep-sea exploration. Deep-sea scholars are regularly confronted with plastic waste, milk cartons, and other human trash. A study of amphipods collected in the Mariana Trench, the deepest spot on Earth, showed the tiny animals were laden with human-produced polychlorinated biphenyls, chemicals banned for decades.⁶ The same study found that every one of the amphipods collected had ingested plastic. Scientists in Northern California spent years anxiously searching for drums of radioactive waste dropped decades ago and rediscovered in the 1990s, while in Southern California, an expedition in 2011 uncovered a debris field with more than 27,000 barrels of the pesticide dichlorodiphenyltrichloroethane (DDT).⁷

Luiz Rocha, a marine scientist at the California Academy of Sciences, specializes in a rare form of deep-sea diving. Most scuba divers are limited by their air tanks to maximum depths of 130 feet. Below that, the crushing pressure compresses the oxygen in the tank and delivers it in such concentrated intakes that it is poisonous. The nitrogen normally mixed into the tank to mimic our atmosphere becomes narcotic below ninety feet.[8] Rocha and a handful of colleagues have pioneered the use of a nitrogen-oxygen-helium blend that allows them to keep breathing as deep as 500 feet, opening entirely new worlds to human visits. These places are so extreme and so little explored that they turn up new species on almost every dive.

In 2018, Rocha was exploring what are called mesophotic coral habitats—corals typically found in the tropics, between 100 and 500 feet, where there may be some light but less than at the surface—off Saint Peter and Saint Paul Archipelago, a remote Atlantic island chain roughly halfway between northern Brazil and West Africa. Four hundred and twenty feet below the surface, on the deepest blue edge of total darkness, Rocha spotted a dazzling reef fish, which he knew was undescribed by science.[9] While his safety diver Mauritius Valente Bell filmed, Rocha closed in on the fish, collector's net in hand.[10] The next moment on film eventually went viral, spawning headlines like this one from the UK *Independent*: "Divers 'Enchanted' by Colourful New Species of Reef Fish Fail to Spot Huge Shark Cruising Above Them."[11] As Rocha pries at the crevice with his net, Bell cries, his voice tinny with helium, "Shark!" But Rocha, focused on the fish, thinks he hears Bell saying to go up. He decides to ignore him. Rocha's head stays down. Bell's voice gets louder and slightly more urgent as his camera swings around to a twelve-foot long sixgill shark. "Look at the shark!" Rocha still doesn't see it. Thirty seconds pass as the shark circles closer and closer. Finally, Rocha swings around and the shark pauses, right over his head, its toothy mouth caught in the glare of the diver's headlight. For a moment, they hang, suspended, looking at each other. Then the sixgill shark turns and slowly swims away. The divers dissolve in high-pitched helium giggles.

It's an epic made-for-the-internet moment that's been watched more than 200,000 times on YouTube, and the fish—Rocha collected it—indeed represented an entirely new genus of reef fishes. But Rocha told us this story and showed us the video to point out something else. As the shark hovers over his head, you get a clear view of the long fishing line trailing from its mouth. Here, in the most isolated islands in the Atlantic Ocean, 420 feet below the surface of a rock inhabited by four people, in the mouth of a shark species that rarely swims within 300 feet of the surface, Rocha had found another human footprint. When he set out for the archipelago, Rocha told us, "I fully expected to find the most pristine deep reefs I've ever seen. Which was not the case. We found trash. There's four people on that island. Every dive we found trash."

Rocha wanted to show us his viral moment with a shark, he said, because "it's one of the saddest things I've ever seen." We are still in the first and earliest stages of our relationship with the deep ocean, a nascent understanding built on only a blink of geologic time, but the relationship already follows some of the contours of our previous handiwork. "Our presence is being felt in places that we've never been," says Erik Cordes, a deep-sea biologist at Temple University. "And that is happening over and over again."

In July 2022, Cordes descended in a small submersible to the deepest part of the Atlantic Ocean. The submersible was visiting the very center of the Puerto Rico Trench, and the scientists and pilot inside couldn't see the rocky walls of the trench or much of anything except mud. But as they reached the seafloor, 20,000 feet below the surface, they looked out the windows and saw a pop-top Pepsi can and a moonshine jar. A dive like that "gives you a sense, physically, of how big and how deep the ocean is," Cordes told us. "But then to do that and feel so disconnected from the surface and so far away and then see trash and a Pepsi can on the seafloor—there's nothing on this earth that we haven't already touched."

And it is not just our trash. The deep sea accounts for over 90 percent of the total volume of the world's oceans—a reservoir so vast that it may seem impossible to change it. Yet there's evidence now that even the deep sea feels the effects of global warming and ocean acidification. And if it is hard to change the deep sea in one direction, it is equally hard to change it back.

As scuba divers the world over know, light disappears rapidly as you sink beneath the surface. By fifty feet, the water has scattered much of the long wavelength red light. By 100 feet, you lose yellow, and by 164 feet, you lose orange, leaving a seascape of deep blues, greens, and violets. Although few divers ever venture deeper, some like Luiz Rocha push into the liminal world in which blue light alone remains. Only the faintest hints of sunlight make it more than 600 feet into the ocean and illuminate a layer that extends another 2,000 feet, which oceanographers label the twilight zone. Below 3,000 feet, there is no light at all, no influence of the sun's heat or regular mixing with warmer surface waters: just the near-freezing black darkness of the deep. Consider again just how new it is for humans to have firsthand knowledge about anything below the shallowest, brightest top layer of the ocean. Astronomers observed the moon through telescopes for centuries before the first human ever set eyes on the deep sea. For the vast majority of human history, anything below the sunlit surface zone has been so far away, so remote, and so impossible to imagine visiting that it stands across cultures for the eternal and unknowable. The word "abyss" itself comes from the Greek for "bottomless."

When research divers started to explore the deep sea in the Papahānaumokuakea Marine National Monument, a vast area in the northwestern part of the Hawaiian archipelago, NOAA consulted with Hawaiian scholar Dr. Huihui Kanahele-Mossman to write a *mele*, a chant requesting permission to enter the deep sea.[12] The mele establishes a relationship between the researcher and the place they want to visit, Kanahele-Mossman explained in a 2022 NOAA webinar. "The sacredness of that area has to be recognized," she said. "Especially in a deep dive. Because we as man, or kanaka, we don't belong there. That is not our place to be." Before every ROV dive into the monument, scientists request permission from the ocean god Kanaloa to enter "the dark blue place of the Sperm Whale, a small piece for me so that I can receive the knowledge of the deep."[13] In *Twenty Thousand Leagues Under the Sea*, which first appeared as a serial in 1869, Jules Verne's scientist narrator has just published two volumes on the mysteries of the deep ocean when he is called for his opinion on the dramatic near-sinking of a ship. (He declares it to be the work of a 300- to 600-foot-long narwhal.) "The great depths of the ocean are entirely unknown to us," he writes. "Soundings cannot reach them. What passes in those remote depths—what beings live, or can live, twelve or fifteen miles beneath the surface of the waters—what is the organisation of these animals, we can scarcely conjecture."[14]

Even as Verne wrote those lines, many Western scientists had only just started to believe the deep sea to be capable of supporting life. Through much of the 1800s, although they had no way to see into the depths or to understand the vastness of the ocean, coastal surveyors had started to drop lines to touch the bottom in places along the coast and to realize the staggering depths the ocean could reach. They had calculated the physics of light and temperature to realize that the deep sea must be cold and dark, and the pressure immense. Because it was difficult to believe anything lived in such conditions, some scientists imagined a vast lifeless world of rocks and mud. In 1843, Edward Forbes, an influential scientist who had surveyed the Aegean Sea and traveled widely in the 1830s and 1840s, popularized the idea that the ocean became increasingly lifeless at depth, a theory he called the azoic hypothesis. For twenty-five years, the scientific establishment argued about deep-sea marine biology on the basis of limited experience and insufficient evidence. Forbes was destined to become a cautionary tale, a pair of British historians wrote in 2006, "of scientists ignoring observations that ran counter to their deep-seated, yet entirely erroneous, beliefs."[15]

While the scientists debated, commercial endeavors on both sides of the Atlantic took an increasing interest in the deep sea. Nearly a century before any human eyes would see the lightless depths, American and British entrepreneurs hatched a plan to cross the Atlantic with telegraph wires.[16] In 1857, an American steam frigate departed the Irish coast carrying 2,500 nautical miles of copper-and-hemp cable, about half the total length necessary, with a plan to meet an English ship carrying the other half of the cable in the middle of the Atlantic. The USS *Niagara* made it several days into the expedition before the cable broke in the open ocean.[17] One of the first transoceanic deep-sea commercial endeavors ended in failure with 380 miles of cable left behind on the seafloor. The next trip succeeded for just long enough to allow the queen of England to tap a message to the president of the United States and send it via a sixteen-hour journey across the depths of the Atlantic—although that cable too soon failed. Investors tried to recover what they could and left the rest behind.[18] The technology kept improving, and the commercial importance of telegraph wires, and eventually of other communications cables, gave the scientific arguments about the nature of the deep sea a stake outside the theoretical. Limited knowledge was an obstacle to global connection: better science could therefore satisfy curiosity and also make the world a smaller place.

In 1869, the same year Verne's *Twenty Thousand Leagues Under the Sea* first stoked European imaginations with its wild depictions of undersea life, the English Royal Society sent a surveying ship called the HMS *Porcupine* to dredge the seafloor off the Irish coast. From among the creatures scraped into a net in 14,610 feet of water, the scientists on board the *Porcupine* pulled two live clams, a mussel, a tusk shell, and a cardiomya bivalve.[19] In a few months of surveying, the expedition found fifty-six new species of mollusks alone, one-fourth of the total known to be living in British seas at the time. English scientists were exhilarated by the conclusive proof of life in the deep sea. Three years after the *Porcupine* expedition, an English ship called *Challenger* embarked on a four-year mission to explore the world's oceans. Visiting every continent except Antarctica, the crew of the *Challenger* sounded, dredged, collected, dissected, and described. The ship's guns were discarded to make room for scientific equipment and dredging winches, and it featured an onboard chemical laboratory. It was likely the first scientific expedition to have both a photographer and sketch artist on board to capture images of the expedition in real time. The voyage is often considered the dawn of the science of oceanography.[20]

As they sailed on the *Challenger*, scientists measured seawater temperatures, density, and chemistry.[21] They collected and named nearly 5,000 species and produced more than fifty volumes of dense descriptive work based on the expedition, ranging from physical and chemical descriptions of the ocean to penguin anatomy, to speculation about sponges they had dragged up from the depths. The line drawings in the fifty-volume set are, to this day, considered works of art. The spindly legs of a crab are drawn outstretched, with hair-like features on each of the ten legs beautifully memorialized, along with knob-like structures on its carapace. The crab drawing notes a preliminary identification of the species (*Amanthia carpenteri*) and that, as depicted, it is "one and a half the natural size." Another page features the intricate structures of numerous species of radiolaria, a microscopic zooplankton that constructs a skeleton of delicate silica rods. Flip to another page and an octopus will stare back at you, tentacles like ribbons flowing across the top half of the page, with each sucker meticulously drawn. It is difficult today for casual fans of the *Challenger* to find complete fifty-volume sets; most of those that were printed went to institutional libraries, and many other sets ended up split apart, becoming collector's items.[22]

Among its many accomplishments, the *Challenger* took systematic soundings of the global seafloor, the first direct evidence of the shape of the deep ocean. In the Pacific just north of Saipan, one depth sounding shot down to 26,850 feet. Later scientists would confirm the Mariana Trench as the deepest place in the oceans, and its maximum extent, 35,876 feet below the surface, is now called the Challenger Deep. In the Atlantic Ocean, the *Challenger* ran into a different kind of depth finding. While sailing between Tenerife in the Canary Islands and St. Thomas in the Caribbean, the scientists noted that the depth dropped off the coasts and then seemed to rise again in the center of the ocean. This relatively shallow area, a seeming underwater mountain range, extended from the North Atlantic to the South Atlantic. The expedition's lead scientist noted that the temperature on one side of the rise did not match the temperature on the other side, so it also seemed to mark a separation point between the western and eastern basins. For the first time in human history, it was apparent that the entire Atlantic Ocean had a seam down its middle. To explain this seam took another seventy-five years and the next major scientific advancement in our knowledge of the full topography of the oceans.

———◆———

To someone accustomed to opening Google Maps and seeing the topography of the ocean floor, it is still amazing to consider that, until the 1920s, the only way scientists could visualize the bottom of the ocean was by someone going out on a boat, dropping a line, noting the depth and the coordinates, and then going back and putting that one infinitesimally small point on a map. This method allowed scientific navigators to develop increasingly sophisticated maps of shallow coastal areas, but when it came to the open ocean, the *Challenger*'s soundings, and the soundings of all the sailing ships that crossed the ocean before it, amounted to pushpin marks in a near-infinite sea. But military technology developed in World War I finally changed the ability to visualize the seafloor. Following the war U.S. Navy ships patrolled the Atlantic with new sonar that captured a continuous measurement of seafloor depth, accelerating the old method and providing so many of those point-by-point measurements that it became feasible to stitch them into a drawing of an ocean basin. Scientists at the Lamont Geological Laboratory (now the Columbia University Lamont-Doherty Earth Observatory) realized they could take all of the data and draft the first underwater map of the Atlantic Ocean. The job of turning those scattered ship tracks into a topography fell to a new hire named Marie Tharp.

Tharp's father was a soil surveyor for the U.S. Department of Agriculture (USDA). She grew up moving constantly and evaluating the landscape as she went. "I guess I had map-making in my blood," she wrote in an essay in 1999.[23] She didn't intend to pursue cartography and went to college undecided on a major. The avenues she found most interesting were largely closed to women at the time, she later wrote, and she didn't want to be a teacher, secretary, or nurse. When World War II started, however, the country needed petroleum engineers, and the men who had traditionally filled those jobs had gone to war. In 1943, the University of Michigan offered a master's program in geology for women, with a guaranteed job in the oil industry awaiting them on graduation. Tharp took the opportunity, earned her degree, and went to work for Stanolind Oil in Tulsa, Oklahoma. On the side, she attended the University of Tulsa and earned another master's degree, this time in math. In 1948, with two master's degrees, Tharp moved to New York City to look for a job in research. She had learned how to draft while at Michigan, so a researcher at Columbia University hired her to help graduate students map the ocean floor. Because the men who ran the oceanographic research considered women "bad luck" onboard ships, Tharp was unable to go on research cruises until after 1968. The genius of her work was done back in the lab.[24]

Tharp followed a three-step process for turning all the sonar data into a map. She would first plot the ship's course on a line through a flat space marked by latitude and longitude. Then she would take the depth measurements and make a two-dimensional rendering of the seafloor topography viewed in profile. Then she would stitch the two together, creating a three-dimensional drawing of the floor as it followed the line at the surface. The work had to be constantly checked and rechecked, drawn and redrawn. "Eventually," Tharp wrote, "I had a hodge-podge of disjointed and disconnected profiles of sections of the North Atlantic floor." But they still were somewhat at the mercy of technology. Opening the ship's refrigerator would interrupt the echo sounder and yield a spot on the map with no data. In her 1999 essay, Tharp added: "Plotted on a map, the ship's tracks looked like a spider's web, with the rays radiating out from Bermuda, where most of the research vessels took on supplies and water. Sometimes, the tracks zigzagged, as the ships fled from the paths of storms."[25]

For the next few years, Tharp worked on the ocean-floor maps, refining and enhancing her drawing technique and incorporating new and better data. In 1956, she and her research partner Bruce Heezen published the first physiographic seafloor map—a map "if all the water were drained away"—of the Atlantic Ocean.[26] It appeared as an accompaniment to Bell Telephone System's Technical Journal.

Several years before she and Heezen published the first map, Tharp had singled out its most significant feature. It was the Mid-Atlantic Ridge again. Tharp noticed specifically that the seam down the middle of the ocean had a valley of its own, a V-shaped indentation that lined up on every one of the six Atlantic profiles she had started plotting. If you boarded a submarine on the U.S. East Coast and traveled east along the ocean floor toward Europe, the seafloor would first sink off the continental shelf, then run along the deep, flat abyssal plain, then rise toward the mid-ocean ridge that the *Challenger* had observed. But instead of climbing to a peak and then quickly descending on the other side, you would climb the ridge only to behold a vast rift valley. Picture not a perfect cone but a volcano with steep sides and depressed caldera. To Tharp, the implication was clear, and it provided support for the theory of continental drift. At the center of an ocean, lava would well up, splitting the ridge as it pushed the plates apart. Heezen was skeptical. The idea of continental drift was widely unpopular at the time. Tharp recalled him initially rejecting the idea of a mid-oceanic rift valley as "girl talk." She knew she would need more data and set about getting it.

Two new sources arrived in 1952. Lamont acquired a new research vessel with a more precise depth sounder, allowing Tharp to make even more accurate seafloor drawings. At the same time, Heezen hired a recent graduate to plot the epicenters of thousands of earthquakes recorded from the seafloor. The original interest was commercial: earthquakes created turbidity currents that destroyed underwater equipment, and Bell Laboratories wanted advice on where to put new cables. It soon became apparent to Heezen that the earthquake epicenters all fell within the mid-Atlantic ridge's rift valley. He and Tharp worked to extend the map of the underwater ridge following the earthquake line. They noticed that near the northern end of the Indian Ocean the mid-oceanic seam made a sharp westward turn into the Gulf of Aden, and the earthquake zone continued right onto land, into the oldest and best-known rift valley on Earth, the East African Rift. When they compared the topography of the terrestrial rift valleys with what their precision depth sounders turned up 8,000 feet below the surface of the Atlantic, the match was inescapable. And the more data that came in from the boats, the more they followed the earthquakes, and the longer the ridge appeared. It was a feature of the North and South Atlantic, the Indian Ocean, and the eastern Pacific. Jacques Cousteau didn't believe what the maps showed and set off on a cross-Atlantic voyage, dragging a camera below his ship, to prove Tharp and Heezen wrong, but he, too, found the ridge and the valley in its middle.[27]

Everywhere the ridge could be found, so could earthquake epicenters. The mid-Atlantic rift valley was in fact one single, continuous ridge, wrapping, just like the seams on a baseball, around the entire planet. The existence of such seams in the seafloor supported an evolution in scientific thinking about how movement occurred on the Earth's surface—not through passive "drift" but rather via a series of collisions and the creation of plates, building mountain ranges and carving the deepest trenches in the world. Modern scientists recognize the oceanic ridge as the longest single geologic feature on Earth. "I think our maps contributed to a revolution in geological thinking, which in some ways compares to the Copernican revolution," Tharp wrote in 1999.[28] "Scientists and the general public got their first relatively realistic image of a vast part of the planet that they could never see. The maps received wide coverage and were widely circulated. They brought the theory of continental drift within the realm of rational speculation. You could see the worldwide mid-ocean ridge and you could see that it coincided with earthquakes. The borders of the plates took shape, leading rapidly to the more comprehensive theory of plate tectonics."

Heezen and his collaborator Maurice "Doc" Ewing presented the finding at a series of lectures in 1956 and 1957, receiving the bulk of the scientific credit. In the 1960s, Heezen and Ewing got into a fierce disagreement, which ended with Tharp, as the only one of them not tenured, getting fired from Columbia. For the next decade, she continued working on her maps from home. At the height of the Cold War, Soviet scientists who appreciated Tharp and Heezen's work stepped in, supplying their own data for her to illustrate.[29] She completed more basin maps, kept revealing different parts of the ocean as if all the water had been drained. Finally, in 1973, one challenge remained: to create the first map of the seafloor of the entire world. Working with Heezen and an Austrian painter, Tharp tracked down and filled in the data to complete a full picture. In 1977, the World Ocean Floor Map appeared.[30] The same year scientists discovered a deep-sea habitat called hydrothermal vents, which would revolutionize thinking about the ecology and geology of the seafloor, and provide another piece of the puzzle that contributed to the development of the theory of plate tectonics.

It took Tharp twenty-five years to acquire the expertise, knowledge, and data to make the World Ocean Floor Map. It is extraordinary and beautiful, a work of art based on the beauty of our planet. In the twenty-first century, our precision tools allow us to map with much greater accuracy and in a comparative blink of an eye. Tharp became a bit of a scientific hero to oceanographers—but only very late in her life. "I worked in the background for most of my career as a scientist, but I have absolutely no resentments," she wrote. "I thought I was lucky to have a job that was so interesting. Establishing the rift valley and the mid-ocean ridge that went all the way around the world for 40,000 miles—that was something important. You could only do that once. You can't find anything bigger than that, at least on this planet." In announcing her death in 2006, the *New York Times* both hailed her scientific achievements and noted the overdue acknowledgment of her big impact on science, including being honored by the U.S. Library of Congress late in her life.

As technical as Marie Tharp's maps of the ocean look, they were largely works of projection. Tharp used thousands and thousands of individual soundings, but these still revealed only tiny slices of the actual ocean floor. Her genius was, in part, to fill in the rest based on her knowledge of geology. After Heezen's sudden death on a submarine research cruise in 1977, Tharp continued to work on mapping, and slowly the technology improved. Research and naval vessels in the 1990s adopted multibeam sonar, which took soundings from the floor in a wider fan shape instead of the single points from previous sonar. The resolution improved, allowing scientists to distinguish features on the scale of a few tens of meters. Now underwater ROVs drag multibeam sonar arrays behind them, using their ability to fly closer to the seafloor to generate even more nuanced views.

In a 2020 webinar celebrating the one hundredth anniversary of Tharp's birth, Lamont-Doherty senior research scientist Vicki Ferrini reminded watchers that still only a small percentage of the world's oceans had been accurately mapped. If you pull up a map of the planet, she said, "It looks like the ocean is mapped. But the reality is that most of that is predicted, it's not actually measured. It's a model and it can be off by kilometers vertically. So if we told the story with this map where we were highlighting where actually data exists, most of this would be blacked out. You wouldn't see anything."[31]

As with so many other stories of the ocean, we are in a moment where we need that information now. "To save Earth's climate, map the oceans," wrote deep-sea scientist Dawn Wright, the chief scientist at the mapping software technology company Esri, in a 2022 op-ed about the importance of fully exploring the deep.[32] Wright and Ferrini have worked together on a Seabed 2030 initiative, part of the United Nations "decade of the ocean," to map the rest of it. "The payoff stands to be tremendous—for everything from ship navigation to climate modeling," Wright wrote. To start, Wright said in the 2020 webinar with Ferrini, much of humanity's modern internet infrastructure relies on the deep sea. Beginning with the transatlantic telegraph cables in the 1850s, the sophistication and number of oceanic cables has increased. Hundreds of major phone and data cables now cross every major ocean basin. Fifteen cables stretch across the Pacific Ocean, connecting East Asia with North America.[33] Over two dozen run across the Atlantic seafloor from the North American East Coast to Europe, the Caribbean, and South America. The Mediterranean and Red Sea are a mass of cables running between Southern Europe, Africa, Arabia, and India. Owned by major telecommunications and internet companies like AT&T, Google, Meta, Vodafone, Verizon, and China Telecom, these lines undergird the modern world economy. Viewed on a map, they look like so many threads binding the continents together.

We also rely on accurate maps of the deep sea for scoping offshore wind energy, understanding fishing habitat, and weather forecasts. People may underestimate the extent to which the ocean matters in their lives, but the weather is an instant way to connect to it. All of our weather, no matter where we live on Earth, is influenced by the ocean. Changing configurations in ocean temperatures swing the global distribution of storms, floods, and drought. The better we understand the shape of the ocean, Wright says, the better we will be able to predict typhoons, tsunamis, and increasingly dangerous storm surges that can inundate coastal areas around the world. Even with all these reasons, "climate change is the most basic and urgent reason to map the ocean as quickly as possible," Wright wrote.[34] A better understanding of the shape of the ocean will enable better measurement of how heat circulates in the deep sea, giving climate scientists more power to model a changing climate.

As the chief scientist at Esri, Wright has cowritten a series of books about the science of mapping the world. For the cover of the second volume, Wright selected a particular map of the world called the Spilhaus projection.[35] The Spilhaus projection looks something like a turquoise-blue heart, crisscrossed with red and yellow veins. The map centers Antarctica and Australia, and on it, all the oceans connect while the land floats free on the margins. It is a way for cartographers to emphasize the way the ocean flows through our entire world. The main heart-shaped center is the vast expanse of the Pacific, Southern, and Indian oceans. There is an artery leaving the heart on the left—the Atlantic Ocean—which narrows into the Arctic Ocean. There are red veins in the image that form the edges of the volcanic Pacific Rim. And, deep yellow veins trace the route of the mid-ocean ridge through the world's oceans. It is a map that, in an unintentional but striking parallel with the Polynesian concept of a blue highway, shows us the world as a sea of islands instead of islands in a far sea. "All landmasses on Earth, no matter how big, are surrounded by the ocean or seas and could therefore be considered islands," Wright said in an interview with *National Geographic* in 2021. "In this light, there's less of a distinction between islanders and mainlanders. On an ocean planet, we are all islanders."[36]

Wright has talked often about how her interest in the ocean and maps started when she was a child growing up in Hawai'i and acting out buried-treasure scenes from the book *Treasure Island*. By the time she was eight, she says, she knew she wanted to work in the ocean. She studied geology as an undergraduate at Wheaton College, then oceanography in graduate school, where she first came across Marie Tharp's 1977 map of the world ocean. It was "a map to my future," she later said. After graduating, Wright worked as a marine technician on a research vessel for three years, spending six months out of the year at sea.[37] In 1991, she became the first Black woman to dive to the seafloor in the submersible *Alvin*, where she mapped the aftermath of underwater volcano eruptions in the Pacific Ocean. She spent decades in academic research, often as the only Black woman in the room, before taking on her role at Esri. She has spoken and written about her frustration at how few Black scientists work in oceanography, and she is a leader in advocating for open access to science and science communication, and integrating social, natural, and physical sciences. "This is a new day now in terms of how scientists are stepping out of the ivory tower," Wright said in 2020. "We're not just in our own little cubicles working on our own data. If you read [Marie Tharpe's] story in her own words or if you read the biographical accounts of what science was like in her time there's not just the gender dynamics that are going on there but it's how people were unable to or unwilling to share their data.... So that thankfully now is another part of how scientific culture is changing."[38]

While Tharp and colleagues were working at mapping the seafloor, scientists were also improving their ability to observe what was happening at great ocean depths. A series of research cruises in the 1960s and 1970s established that, in the Galapagos Rift, a relatively well-studied area of the deep sea 250 miles northeast of the Galápagos Islands, there were areas of elevated water temperature just above the seafloor as well as unusual mounds and clusters of small earthquakes very near the elevated water temperatures.[39] In 1976, a U.S. Navy ship using narrow-beam sonar captured an accurate map of the rift zone.[40] Oceanographers and engineers at the time focused on improving access to the deep sea via submersibles. And interest in exploring the Galapagos Rift coincided with the development of two new exploration vehicles: *Angus*, a towed sled with cameras and strobe lights, and *Alvin*, a submersible that had been taking scientists and U.S. Navy personnel to the deep sea since the mid-1960s.[41]

In February 1977, a large team of geologists and biochemists from Woods Hole Oceanographic Institution (WHOI) set out on an expedition to use the new technology to see what was causing the unusual deep-sea conditions. The team planned to drag *Angus* via a two-mile-long cable attached to the ship at the surface. A winch operator lowered the *Angus* into position over the seafloor, then raised and lowered it manually as topography required. Transponders dropped separately to the ocean floor helped ping the sled and triangulate its position relative to the bottom. The surface ship slowly motored forward, dragging the *Angus* over the seafloor at three-quarters of a knot, slower than even a slow walking pace. A film camera turned on and started taking pictures, accompanied by a bright strobe light, every ten seconds. As the surface ship slowly carved through the dark sea, a three-person crew kept track of the altitude, the course, and the water temperature as the sled floated through the darkness. Sometime around midnight, about six hours into the watch, the water temperature around the sled suddenly spiked. The anomaly lasted three minutes. Then it cooled down, and the *Angus* drifted on.[42] After twelve hours, the *Angus* ran out of film. It had taken 3,000 photographs, creating a 400-foot-long filmstrip. The scientists developed the 35mm film on board by the next morning, then sat down to review the photos frame-by-frame.

The bottom below them was mostly made up of vast fields of pillow lava, "like mounds of toothpaste," expedition scientist Robert Ballard later wrote. As the timestamp on the frames approached the temperature anomaly, the scientists sat forward. "The photograph taken just seconds before the temperature anomaly showed only barren, fresh-looking lava terrain," Ballard continued. "But for thirteen frames (the length of the anomaly) the lava flow was covered with hundreds of white clams and brown mussel shells. This dense accumulation, never seen before in the deep sea, quickly appeared through a cloud of misty blue water and then disappeared from view. For the remaining 1,500 pictures, the bottom was once again barren of life."

On board the ship, scientists were already preparing for another dive, this time with *Alvin*. Two scientists and a pilot climbed into the submersible and dove to the site of the anomaly. As they approached, Ballard wrote, shimmering clouds of superheated water rose through cracks in the lava. As the hot water hit the freezing cold water of the depths, minerals like manganese precipitated out, coating the seafloor and creating chimney like structures. Even more surprising, the area around the vents was covered with life.

For the first time, scientists were looking at life based on an energy source other than photosynthesis. Giant tube worms, several feet long with a bright red plume of color extruding from the tubes, covered the active vent areas. The tube worms host microbes, which oxidize hydrogen sulfide in the superheated and mineral-rich water that flows through cracks, fissures, and chimneys at the seafloor. Crabs, fish, and shrimp live off this chemosynthetic food web, sometimes consuming microbial mats that form on the seafloor, other times picking at the flesh of the tube worms themselves. The water flowing out of the vents comes out scalding, prevented from boiling by the intense pressure exerted by thousands of feet of overlying water. Organisms orient around the sites based on their heat tolerance.

The revelation that life existed at the deep-sea vents thrilled scientists, revolutionized the field of biology, and set off searches for similar communities elsewhere. Hydrothermal vents are now documented around the world, and differences in the ecology, mineralogy, and geochemistry of vent systems continue to be discovered. Chimneys have been documented that extend hundreds of feet off the seafloor. Decades after the *Alvin* expedition, biologists returned to the Galapagos Rift and found that one vent, named the Rose Garden, had been completely paved over by a lava flow. A few hundred feet away, a new hydrothermal vent community had formed. By watching these sites through time, scientists have been able to document a progression of species as vents change in temperature, age, and geochemistry, much as there is a succession of species in a constantly shifting forest, tide pool, or marsh.[43]

The hydrothermal vent discovery also helped establish *Alvin* as a critical tool in the new science of the deep sea. "I don't think anything is as powerful as being there yourself," Temple University biologist Erik Cordes told us. "You're in a vehicle physically moving through a three-dimensional space. You're not watching it on TV." Video, samples, and data collected from *Alvin* over the last half-century are responsible for much of the knowledge that we have about deep-sea environments, including vent ecosystems, underwater volcanoes, deep rift valleys, midwater fish, decaying whale carcasses, and bubbling methane seeps on the seafloor.[44] Starting in 2011, *Alvin* was overhauled, part by part, with new materials that now let it reach depths of 21,000 feet—making 98 percent of the seafloor accessible.[45] Once the submersible was certified in 2022, Cordes was part of a scientific team to test its new capacity on a dive to an unexplored part of the Puerto Rico Trench. He told us that he had been on forty-six dives in submersibles and remembers every one, and this one was memorable in part for the incredible depth—more than twice as deep as he had ever gone before. *Alvin* collected a black coral from below 20,000 feet, one of the deepest ever recovered from the Atlantic, and some plastic for a study of what might live on deep-sea plastic. Cordes saw isopods that he had read about but never seen before and thought they might be a new species. "The experience of being that deep, and thinking about 6 kilometers of water over your head while you're on the bottom of the ocean was pretty spectacular," Cordes said.

Alvin holds three people on a dive, generally two observers and a pilot. The pilot maneuvers the submersible and monitors the life support systems. The scientists run the cameras and try to document everything they see outside. A dive might spend three hours on the bottom, and it goes in the blink of an eye, Cordes told us. "You watch the video and it's the sub[mersible] going very slowly over the muddy sea floor, but when you're in there it's panic," he said. "Sitting around trying to do six things at once. Feverishly sitting."

In 1983, marine geologist Charlie Paull and several colleagues dove in *Alvin* and found a new chemosynthetic environment in the Gulf of Mexico, where a biological community similar to what scientists had seen around hydrothermal vents seemed to live off sulfide-rich water bubbling from the seafloor.[46] Unlike the vents, however, these areas didn't show any sign of elevated temperature, leading to their name, cold-seep environments. Microbes oxidize hydrocarbons or sulfur-rich seepage from the seafloor and form the base of the food web populated by mussels, clams, shrimp, and fish. Paull and colleagues explored these environments in six dives on the submersible *Alvin* and wrote in *Science* in 1984 that cold seeps hosting chemosynthetic communities could be common in regions where fluid or gas seeps out along the seafloor. Since that first discovery, cold seeps have been found around the world in a wide variety of conditions.

Scientists have also documented deep-sea pools of brine—lakes of dense water at the seafloor that form due to high concentrations of salt seeping out of rocks below. The pools tend to be associated with seeping methane and hydrogen sulfide and are low in oxygen, making them inhospitable to most life, but they are often rimmed with chemosynthetic communities where the salty, mineral-rich water meets more normal marine waters.[47] Waves lap at the shoreline of the brine lake, and larger waves that form due to tectonic movements or nearby underwater landslides may cause the salty, anoxic waters to spill out into the surrounding environment, killing nearby marine life.[48]

Chemosynthetic communities can even form around an unlikely source: the skeletons of dead whales. Whale falls produce a rapid export of carbon from the surface to the deep sea. In 1987, scientists diving in *Alvin* found the intact skeleton of a sixty-five-foot-long whale in 4,000 feet of water in the Santa Catalina Basin off Southern California. All around the whale fall site, the scientists found mussels and clams that appeared related to those found at cold-seep and hydrothermal vent sites.[49] In the years since, numerous whale falls have been documented and studied, and scientists have even sunk dead whales in the deep sea to watch what happens. At first, larger predators such as sharks and hagfish remove the whale carcass flesh. Smaller crustaceans and worms then eat away at the remaining particles of organic matter left on the bones. Finally, microbes begin to feast on decaying bone, which produces hydrogen sulfide. This phase enables microbial mats and bivalves that harbor chemotrophic endosymbionts—just as in hydrothermal vents and cold seeps—on, within, and in a halo around the whale carcass that can last for years, or even decades to centuries.[50] Whale falls might be the most biodiverse habitat of the deep sea, exceeding even hydrothermal vents.[51]

Even before finding whale falls, scientists suspected that large deposits of organic material in the deep sea might be able to support unique ecosystems. Decades of research on decaying wood in the ocean had documented the prevalence of wood-eating "shipworms." Wood-eating invertebrates are well studied because of interest in the decay of wood structures in the ocean, including both shallow- and deep-water shipwrecks. While not all wood falls produce chemosynthetic habitats as part of their decay, microbial communities similar to those documented at whale falls have been observed with deep-sea wood deposits, which are often trunks or parts of trees that have been deposited in deep water after storms or similar events.[52] Studies of deep-sea wood-boring bivalves were among the early experiments facilitated by the *Alvin* submersible. Scientist Ruth Turner spent her career documenting and investigating wood-boring invertebrates, and in 1971, she became one of the first women to dive in *Alvin*. In 1972, she began a series of experiments where she pushed panels of wood into the sediment in 6,000 feet of water and left them there for 104 days. When she returned, she saw the wood overrun by bivalves.[53] Turner spent the rest of her career studying wood-boring invertebrate communities, including on the *Titanic* shipwreck.[54]

After decades of exploration of these deep-sea chemosynthetic habitats, scientists know a lot about where they may be located, how they might be similar, and how species may have evolved to live in these environments. Recent work has even attempted to understand how these environments may be connected—environmentally and genetically. Vents, seeps, and falls all start with microbes that can turn reduced inorganic matter into carbon and kick-start a food web. And while the life that arises differs dramatically from place to place, several scientists wrote in 2012, "together, these ecosystems create a network, extending along margins and across ocean basins, of soft-sediment habitats fueled, at least in part, by the oxidation of reduced chemicals."[55]

In 1962, just a few years after Marie Tharp had generated the first map of the Atlantic Ocean floor—a map that did not yet include the Gulf of Mexico—a drilling ship operated by the Shell Oil Company took core samples from the Gulf in several hundred feet of water and found oil. Thousands of feet below the surface, buried under water and mud, the ancient remains of plants and animals from a highly productive inland sea slowly cooked over hundreds of millions of years into some of the world's largest fossil fuel deposits. As the Mississippi River transported mud into the Gulf, the crushing pressure of all that mud building up over time caused layers of sediment and salt deposits to rise toward the seafloor in massive domes, creating a pockmarked ocean bottom that effectively concentrated and captured hydrocarbons in reservoirs.

Offshore oil drilling had long outpaced scientific understanding of the seafloor or efforts to regulate it. In 1896, while scientists were still making sense of the reports from the deep-sea explorations of the HMS *Challenger*, the first oil well was constructed over the ocean off wooden piers in California.[56] Shortly after World War II, the first offshore platforms, completely out of sight of land, started operating. Three years after Rachel Carson wrote in the *Sea Around Us* that the deep sea "has withheld its secrets more obstinately than any other," the first mobile offshore platform, a submersible drilling barge called Mr. Charlie, drilled its first exploratory well near the mouth of the Mississippi River. The potential depths for a well kept increasing: 600 feet in the 1960s, 1,000 feet in the 1970s, 10,000 feet in the 1990s. These discoveries were announced in shareholder meetings and in oil company prospectuses and celebrated at annual engineering conferences.[57] In 1996, Shell began operating from a platform it called Mars—a platform on which it spent three times more than NASA spent to launch, in the same year, the *Pathfinder* rover to the planet Mars. Deep-sea scientists today still talk sometimes about how we know more about the planet Mars than we do about the deep sea, but we have long known more about—and invested more in—oil exploration than we do about Mars *or* the deep sea.

After a catastrophic oil spill off Santa Barbara, California, in 1969 slicked popular beaches and devastated wildlife, environmentalists in California and on the Atlantic coast pushed successfully to ban offshore drilling. Yet the country still demanded oil, and after price shocks in the Middle East, wanted it produced domestically. Stringent new environmental regulations applying to leases and well development exempted the Gulf of Mexico and thus promoted drilling there. In the 1980s, the U.S. Department of the Interior broadcast its intention to accelerate and expand oil production in the Gulf of Mexico by selling leases to 1 billion acres offshore.[58]

In 2008, the government sold a lease for a nine-square-mile part of the Mississippi Canyon to the oil giant BP for $34 million. A massive deepwater rig started drilling a year later in Mississippi Canyon Block 252 at a prospect site they named Macondo, after the fictional town in the Gabriel García Márquez novel *One Hundred Years of Solitude*. BP had never drilled in the area before, and it did not have much information about the geology below the seafloor. The company planned an exploratory well to figure it out. From a surface rig floating in 5,000 feet of water, it would drop a pipe down to the bottom, then drill another roughly 15,000 feet into the seafloor rock and mud.[59] Only once the hole was dug would they be able to tell if there was a valuable reservoir of oil worth tapping. After its first rig was damaged by Hurricane Ida, BP's drilling contractor sent another rig to the site to finish the job: the Deepwater Horizon.

Beyond cold seeps and brine pools, deep-water corals form one of the most critical habitats in the Gulf of Mexico. (There is still some debate over exactly what constitutes a deep-sea coral reef, Erik Cordes told us. A "reef" was traditionally defined as a structure that is hazardous to ships. Because deep-sea coral structures aren't a hazard to surface ships but otherwise resemble in many ways their shallow-water counterparts, what should they be called?) Oil companies began documenting the presence of deep reefs in the 1980s in the North Sea, likely because the companies had ready access to submersible and ROV technologies that were just starting to be used by scientists.[60]

Unlike the life in chemosynthetic vent communities, deep-sea corals need the surface. They primarily feed on marine snow, the organic detritus that drifts constantly downward from the sunlit and highly productive surface. Fish, crabs, anemones, and echinoderms live on and around the coral. Corals grow slowly—many over tens or hundreds of years. Isotopic analyses have identified colonies of black corals in the Mediterranean and around Hawai'i that are more than 4,000 years old.[61] These habitat-forming corals can expand to cover huge areas of the seafloor, even over a mile or more, which is why they are called reefs like their shallow-water counterparts.[62] As they grow and expand, so does the life around them.

The field of deep-sea coral science is a young one. *Lophelia pertusa*, probably the most well-known deep-sea coral, has been studied closely only since the 1990s in the North Atlantic. The first International Symposium on Deep-Sea Corals was held in 2000, and the submersible *Alvin* was first used to sample deep-sea corals on seamounts in the North Atlantic in 2003.[63] Deep-sea corals come in an enormous variety of shapes: giant translucent fans that shimmer in the beam of the submersibles, ten-foot tall bubblegum corals that stand like a forest of bright pink broccoli, feathery fronds that sway in the current like palm trees catching the trade winds, and sometimes single stalks bending in the ocean current. Exploratory dives like those in other deep-sea habitats regularly result in the discovery of new species even in otherwise well-known parts of the ocean.

In the Gulf of Mexico, scientists have documented reefs for decades in the medium-depth mesophotic zone, where some light reaches, and in the depths completely devoid of light. One of the earlier sites to be studied and protected, Flower Garden Banks in the northwestern Gulf of Mexico, sits at a range of 56 to 722 feet on and around underwater mounds formed by salt domes in the subsurface rocks. Although the waters are colder and darker than typical surface coral reefs, scientists have documented a wide range of species, including shallow-water, reef-building coral species and associated life—coralline algae, fish, sponges, shrimp, crabs, and sea stars—toward the shallower tops of the mounds. While some parts of the Flower Garden Banks are accessible by traditional scuba, the remaining habitats have been documented and explored by technical divers and ROVs. These deeper areas are home to a range of mesophotic reef species and extend all the way into full darkness with deep-sea coral species.[64] Scientists began documenting the diversity of these shallow, mesophotic, and deep reef environments in the Flower Garden Banks and spent about twenty years building a case for protection of this environment—with potential damage from oil drilling front and center in their minds—before it was designated as the tenth U.S. National Marine Sanctuary in 1992.[65] Beyond the Flower Garden Banks, deep-sea corals are widely distributed on the seafloor of the Gulf of Mexico. An atlas of known locations of deep coral habitats compiled by NOAA shows a Gulf that is dotted with sites of gorgonians, black corals, stony corals, sea pens, and sponges.[66]

Most animals that live in extreme cold grow and mature slowly. Slowness is an adaptation for dealing with the frigid, lightless depths and with potentially limited food availability. Yet it is a poor adaptation for dealing with human wrecking balls, whether in the form of bottom-scraping fishing trawls or other disturbances. Scientists have compared the potential effect of fishing trawls on deep-sea corals to the damage from clear-cutting a forest.[67] What has taken hundreds of years to build, mature, and diversify can be wiped out quickly. Scientists who spend time in the deep sea worry, considerably and with good reason, about what humans might do—and have already done—to it. In op-eds, YouTube talks, and peer-reviewed science journals, they argue for applying the precautionary principle to our relationship with the deep ocean. "It takes a little bit of forethought, which is something that sometimes we're a bit short on," Cordes says. "But we can do this. We have the capacity to preserve these pristine environments before we've had an impact."[68]

Between January and April 2010, the Deepwater Horizon drilling ship kept extending the well at Macondo. By April, it had reached 18,360 feet below the surface.[69] Engineers concluded it would not be safe to go any deeper, but it soon became clear they would not need to. Studies of the well over the next few days showed that they had tapped into at least 50 million barrels of oil. As a rig drills, it layers steel casing to reinforce the well, and now BP ordered the Deepwater Horizon to fill cement behind the well's steel walls to stabilize it and then close it off so that a less-specialized rig could come out to collect the oil. On the morning of April 20, 2010, workers on the Deepwater Horizon started and finished the process of pouring cement in a hole 18,000 feet below the seafloor.[70]

Throughout the day, the job seemed to be going well. But by the evening, anomalous pressure tests showed that something wasn't right. It was a busy time on the rig, but engineers seemingly dismissed the test results. Around 9:40 p.m., mud from the drilling rig spurted out and covered the platform. The mud was a sign that the light hydrocarbons from the oil reservoir had pushed into the well and were racing upward, displacing the mud as they went. The engineers realized the alarm and tried to shut off the well, but it was too late. In scuba class, divers are taught that as they rise toward the surface, the pressure decreases, allowing gas to expand and rise even faster. When they are diving, they vent gas from their buoyancy vest as they rise, deflating it to control the speed of their ascent. The same physical process governs gas under immense pressure deep beneath the ocean floor. If you are floating on top of an 18,000-foot-tall straw that is connected to a 50-million-barrel reservoir of lightweight hydrocarbons, you must avoid the gas entering the straw and rising through it.

On April 20, 2010, on the Deepwater Horizon, the safety measures failed. By the time the mud splashed onto the deck, gas was already rocketing up the drill pipes, expanding and accelerating as it rose. At a congressional hearing a few months later, a representative from Transocean (the company that owned the Deepwater Horizon itself) compared it to a 550-ton freight train, with a "jet engine's worth of gas."[71] At 9:49 p.m., the first explosion rocked the rig. Employees were thrown across the room, trapped in debris and wreckage. Power and communications went down, and the rig went dark. Eleven people died. Hundreds of workers, many badly injured, managed to evacuate into lifeboats.[72] Twenty-seven hours later, at 1 a.m. on Earth Day, the survivors of the worst offshore oil disaster in U.S. history made it back to land.

Within a few weeks, ROV imaging showed a dramatic oil plume spurting from the broken well. BP tried and failed to close it off. The company's only clear fallback option was a monthslong process to dig a second well into the first to relieve the pressure on the blown-out well and then cap the new well to stop the flow. Thousands of gallons of oil were now gushing into the Gulf of Mexico every minute from the deep seafloor, with no obvious way to stop it.[73] The gushing rapidly became the nation's largest oil spill disaster. Federal, state, and local officials flew in; tens of thousands of people worked to shut off the Macondo well, mitigate its damage as oil flowed toward coastal Louisiana, and clean up oil as it washed on beaches and pooled at the surface of the water.

Journalists, scientists, and members of the public criticized what seemed like a lack of planning for an emergency like this. In an article a year later in the *New Yorker*, journalist Raffi Khatchadourian wrote that there was a disaster-response plan drawn from what the government had learned from previous oil disasters and from managing firestorms in California in the 1970s.[74] But if there was a plan for a government response to a real-time and ongoing environmental calamity, it was not a match for the catastrophe at the Macondo well. By the time the Macondo well blew out, there was no simple, effective action that any individual politician, manager, or scientist could take—short of inventing a time machine. When it came to the Gulf of Mexico, humans had charged into the deep sea without full information and without a risk assessment—in fact without the ability to assess the risks accurately.

Scientists had trouble predicting where the oil would flow because they still lacked basic oceanographic information about the region.[75] BP's risk assessment, submitted in advance and approved by regulators, had been copied and pasted from a NOAA website and referenced Alaskan animals like walruses and sea otters, neither of which lives anywhere near the Gulf of Mexico. The ecological expert the company had named as its first responder in case of a spill had been dead for several years.[76] Proposals to cap the well had to be evaluated against the risk of losing the rest of a vast reservoir of oil that Americans still intended to burn someday.[77]

The damage being done was, as government investigators later wrote, an inevitability that was decades in the making. It mirrored wildfires as well as coral bleaching, fisheries collapses, and ice sheets melting. Each disaster had happened as the consequence of decades or centuries of management decisions. By the time the Macondo well was capped three months after it collapsed, 5 million barrels of oil had surged into the Gulf of Mexico.[78] Humankind's first visit to Mississippi Canyon Block 252 and the Macondo drill site ended the way most of our first visits to new places end: with our fingerprints left behind at the scene.

In 2009, a year before the Deepwater Horizon disaster, Cordes and his research group cruised by the well site. They dove at a site about six miles away and spotted a few corals, but they did not stop for long. One year later, that became the extent of what was known about the deepwater corals in the near vicinity of the Macondo well. Slightly farther away, researchers had identified and visited several coral sites within fifty miles of the well, and before the well exploded, they had already scheduled cruises to revisit them in the fall of 2010. Those cruises became the best opportunity to look for damage from the spill.

In an oil lease block twenty-three miles from Macondo known as Viosca Knoll 906, ROVs had found a large reef of the slow-growing, long-lived coral *Lophelia pertusa*.[79] When scientists revisited Viosca Knoll in the months after the spill, they found the coral intact and thriving, with no apparent harm. When the ROVs explored closer to the well site, they discovered a previously unknown coral community seven miles southwest of Macondo, at a site called Mississippi Canyon Block 294. These corals, mainly branching gorgonians called *Paramuricea*, were covered in a thick brown goo known as flocculent, or floc. Samples taken from the goo showed that it contained both oil from the blowout and a chemical from the dispersants that had been used to try to clean up the spill.[80] Compared to corals in areas the spill had not reached, their branches were less healthy.[81] Hydroids had colonized many of the corals. Based on previous expeditions that had built up knowledge of what a thriving deep-sea coral community looks like in the Gulf of Mexico, scientists knew this group was sick. "We know because we had been exploring the deep sea in the Gulf of Mexico what those deep-sea ecosystems were supposed to look like," Cordes said. "We know where the damage from the oil spill occurred, even in the deep sea. But only because we had been down there. Only because we had been exploring."[82]

Researchers kept coming back to the Mississippi Canyon Block 294 site to photograph the corals every year from 2011 to 2017. On their last survey, seven years after the blowout, the corals still had not recovered. "Colonies with more extensive injury may never recover," scientists documented in 2018.[83] Even many miles away from the well site, corals in the path of the oil plume suffered. At a series of shallower sites near the Louisiana-Florida border, researchers found numerous corals with damaged or lost branches. Those, too, did not recover over the length of the study.[84]

Although there had been scientific surveys of deepwater habitats in the Gulf of Mexico starting in the 1970s, there was little baseline environmental information for the area within fifty miles of the Macondo well. Two prior trawl studies "were unfortunately conducted too far to the east and west of the spill site to provide a description of baseline conditions at the impact site," marine scientist Marla Valentine later wrote. "Without detailed information pertaining to the organisms inhabiting this area, it is difficult, verging on nearly impossible, to determine the impacts of this oil spill and the potential resiliency of these inhabitants."[85] Engineers on the Deepwater Horizon had twice flown an ROV around the bottom near the well site in the months before the spill to look for life, although they did not use a navigation system.

After the oil spill, research funds and settlement money from BP started flowing in to fund research about the Gulf of Mexico and the ecosystems around Mississippi Canyon Block 252. In the area near the well site, researchers soon documented declines in numerous small seafloor dwelling organisms, including foraminifera, copepods, and small invertebrates called meiofauna.[86] In August and September 2010, one month after the oil had stopped flowing, Louisiana State University (LSU) researchers Valentine and Mark Benfield used ROVs to establish transects in Mississippi Canyon Block 252 for the first time. "This study is to our knowledge, the earliest comprehensive attempt to characterize biological conditions around the Macondo well," they later wrote.[87]

It is the same process as that used to document tide pools and coral reefs: drive the ROV a set distance between two fixed points, and count and describe everything you see in the video. And even after the spill, there was a lot to see. The ROV filmed crabs, sea stars, anemones, sponges, sea cucumbers, corals, and dozens of species of fish, including several eels and a large sixgill shark.[88] Valentine, now a fisheries scientist at Oceana, submitted the project as her master's thesis at LSU. The thesis states repeatedly that it is hard to say the exact effect of the oil.[89] When the well exploded, the oil plume lifted off the seafloor and flowed to the south, and the southern transects generally seemed to show less abundance and less diversity. Valentine assessed four transects 2,000 meters from the blown-out well, and one 500 meters away; the closer site had the least number of species and the least overall number of animals. Why? Based on the absence of surveys beforehand, researchers could only speculate. But Valentine and Benfield's project was something like the philosophical reverse of the Hewatt transect. Where Hewatt established a baseline in an abundant and prewarming world, against which future movement and assumed biodiversity loss could be measured, Valentine and Benfield were establishing a baseline for Macondo, a site next to an oil spill. The transects could be repeated in future years to measure recovery—if it happened.

Without clear baseline information, scientists also tried to infer what the oil had damaged. Young fish that ingest oil suffer heart defects, and the spill poured through the spawning ground of tunas and jacks in Mississippi Canyon Block 252.[90] "Losses of early life stages were therefore likely for Gulf populations of tunas, amberjack, swordfish, billfish, and other large predators that spawned in oiled surface habitats," a group of scientists wrote in a summary report in 2016.[91] A fish survey from the deep sea one year after the spill collected 460 species of fish; fifty of them had never been recorded from the Gulf of Mexico and one was an entirely new species of anglerfish, confirming that the deepwater Gulf is both a biodiverse place and an understudied one, but not offering much insight into how the spill had affected it.[92] Whales, dolphins, seals, and sea lions all swim through the Gulf of Mexico, and scientists had tracked their populations since before the spill, but because no one anticipated a before-and-after science experiment directly around the Macondo well, and most cetaceans die and sink without ever being noticed, it's hard to say how meaningful the die-off was for marine mammals. The U.S. Fish and Wildlife Service directly counted 140 dead animals in the year after the spill, a number that might undercount the true mortality by fifty times.[93]

Most of the oil stayed deep, or ended up returning to the seafloor as a "dirty blizzard" of marine snow, leading scientists to the realization that we were dealing with the effects of essentially two oil spills—one at the surface entangling wildlife and washing up on beaches, and one in the deep sea. Marine snow is normally the slow and consistent rain of particles from the surface, connecting shallow ocean productivity to the deep sea and feeding an array of organisms living in the deep. But here, it became a toxic rain. Oceanographer Uta Passow and colleagues labeled the new phenomenon marine oil snow.[94] Much of the snowing oil appeared never to have seen the surface, Passow and a colleague wrote in 2021—it emerged from the leaking well, glommed onto existing particles in midwater depths, and sunk.[95] A layer nearly half an inch thick of sediment, marine snow, and oil also blew out from the well, nearly instantaneously reaching the deep sea in an area extending over 100 miles away.[96] Marine oil snow continued to drift downward for about a year.[97] In the immediate aftermath of the spill, areas of the seafloor came to "resemble a toxic waste dump," one researcher wrote in 2016.[98]

In 2017, three scientists—Mark Benfield, Clifton Nunnally, and Craig McClain—returned to the well site transects to see what had changed. Their funding for follow-up studies had largely ended in 2014, they wrote, and the study was possible only because McClain had an ongoing research project that allowed them to study control sites. The dives found a totally different environment near the well than Valentine and Benfield had seen right after the spill in 2010. Diversity had decreased in the seven years since the spill, and an army of arthropods had moved in. Three kinds of animals, sea stars, flatworms, and peanut worms, had disappeared from the site entirely, although there were plenty of them present at the control sites.[99] Benfield, Nunnally, and McClain wondered whether oil had attracted the arthropods—crustaceans in particular find it enticing—and poisoned the other species. But as usual, it was hard to say. "The lack of natural history information for deep-sea animals prevents the determination of species with ecological traits that enable them to be opportunistic or pollution tolerant," they wrote. It was obvious, however, that the oil was still around. The researchers reported watching a crab walk through the mud, poking into a clearly oiled layer with each claw print.[100]

In the early days of the research submersible *Alvin*, as scientists were working off the coast of WHOI, the submersible's support cables broke during a launch, and water poured in through an open hatch. The pilot escaped, but *Alvin* sank to the bottom in 5,000 feet of water.[101] Due to weather conditions and the need for specialized equipment to recover the sub, *Alvin* sat at the bottom for nearly a year. Losing *Alvin* at the beginning of such an important age of discovery for the deep sea was a disaster at the time—but also an unintended experiment in deep-sea microbial communities. The crew had lunches packed and sitting inside *Alvin*, including apples and bologna sandwiches wrapped in wax paper. When the recovery effort successfully refloated *Alvin* back to the surface so that needed repairs could take place, a startling discovery was made: the lunches appeared mostly intact.[102] A year later, the apples seemed "pickled" and the bologna was "discolored" but still well-preserved. What scientists learned is that the combination of high pressure, very cold temperatures, and low oxygen conditions at depth dramatically slowed the microbial processes that would have degraded the sandwiches and apples in mere days to weeks at the surface.[103]

There might be thousands or tens of thousands of microbes in a teaspoon of seawater, Bigelow Laboratory for Ocean Science senior research scientist Beth Orcutt said in a 2022 NOAA webinar,[104] or maybe even millions as you approach a high biodiversity area like a deep-sea vent. Deep-sea microbes provide the basis of life for animal communities like the hydrothermal vent tube worms, precipitate minerals that form on the ocean floor, and sequester carbon. They harbor enormous, barely explored genetic diversity, holding the potential keys to new antibiotics, cancer treatments, and more, and yet we know almost nothing about them. Orcutt's lab asks some very basic questions—"Who are they?" "What do they do?" "And how fast?"—because even those questions have not yet been answered. The small fraction of ocean floor off which we have bounced sonar beams, so often mentioned by scientists as proof of how little we have explored the deep sea, absolutely dwarfs the percentage we have directly visited to study its biology.

"The deep ocean includes places that could be as iconic as the Okavango Delta, the Himalayas, and the Grand Canyon, and potentially as important as the Amazon rainforest," marine biologist Diva Amon, the founder and director of the ocean nonprofit SpeSeas, said in a 2022 online question-and-answer session "When we find these places, we need to protect them. But we can't effectively manage and protect what we don't know, understand, and value."[105]

As federal funding for deep-sea research has grown more limited, scientists have increasingly turned to private funders, fueling another revolution in deep-sea exploration. After decades of U.S. scientists depending only on limited research funding and ship time made available by the National Science Foundation (NSF) and NOAA, public-private partnerships sprung up and breathed new life into understanding the deep sea. In 2008, Robert Ballard, well known for his involvement in the discovery of hydrothermal vents and for his 1985 discovery of the wreck of the RMS *Titanic*, founded the Ocean Exploration Trust, which hosts research cruises around the world on its ship the E/V *Nautilus*. These expeditions have partnered with federal agencies, universities, private foundations, and the National Geographic Society to support the science of the deep sea. In 2012, the Schmidt Ocean Institute began ocean expeditions on a state-of-the-art research vessel named the R/V *Falkor*, again in partnership with national and

THE DEEP 203

international foundations, agencies, and universities. Both organizations have advanced the use of live-streaming technology to send images and data back to scientists and to an increasingly interested and engaged public audience.

In a 2022 cruise on the E/V *Nautilus*, scientists explored previously unknown environments in the Papahānaumokuākea Marine National Monument. The expedition focused on the Wentworth Seamounts, a chain of volcanic features that appear slightly offset from the primary Hawaiian island chain. A Native Hawaiian cultural working group nomenclature committee named the expedition Luʻuaeaahikiikapapakū to represent "the journey to and the work in the papakū, or the ocean floor, which includes surveying and mapping seamounts, and investigating macro-biology, corals, and deep-sea rocks in the Wentworth Seamounts."[106]

In live-streamed videos, the strobe lights of the two ROVs deployed from *Nautilus*—the *Argus* and *Hercules*—penetrate the dark and slowly reflect off white sponges and the pastel pinks and yellows of deep-sea corals, perched on blocks of black basalt. The crew identifies four different species of bamboo corals as they come into view, and then the rovers pause over a brightly colored yellow coral, which a scientist on board identifies as an *Acanthogorgia*. Delicate orange brittle stars wind around the branches of the coral, living within the soft polyps. The brittle stars climb up the corals like a ladder, using them to dangle their delicate arms in the flowing currents. The cameras pan across the seafloor, and a pink sea cucumber ambles along. "Fancy cucumber!" one of the crew notes.[107] On the first-ever visit to a 4,000-foot-deep seamount they refer to as Seamount A, the ROVs hover over pillow-like black basalt flows to observe a bamboo coral that the scientists estimate is six to nine feet tall—taller than the ROVs. The coral is a single stalk, waving in the currents, with a light white color and red polyps.[108] The scientists narrate the footage they see from the ROVs, which is transmitted in real time around the world via satellites. Scientists on shore and interested members of the public watch the *Nautilus* dives, posting on social media about what they are seeing. There are also ship-to-shore presentations where scientists involved in the research explain what they are seeing to K–12 and college classrooms, museums, and other public venues.

Beth Orcutt, one of the lead scientists on the Luʻuaeaahikiikapapakū expedition, studies the microbial communities on rock samples, and the valuable mineral crusts that are frequently found on the seafloor. Fluid flow at the seafloor does more than just establish a basis for life in the deep sea. Interactions between seawater and rock also cause minerals to precipitate and, in some places, to coat the seafloor. Copper, nickel, aluminum, manganese, zinc, lithium, and cobalt—the metals that power modern battery-based infrastructure like smartphones, electric cars, solar arrays, and wind turbines—have accumulated over time on the deep seafloor.[109] These minerals are also increasingly hard to find on land because terrestrial mines have proven to be dangerous, environmentally hazardous, and limited in capacity. Understanding the microbial communities and the geochemistry of these rocks will provide important baseline information for inevitable future decisions about the deep sea.

Deep-sea mining technology has now advanced out of the realm of science fiction to the point where it is feasible to gather those precious metals. A mining operation might send down a scraper to the surface of the seafloor and a vacuum to suck it all up. For scientists who have spent the last few decades watching oil and gas exploration proceed faster than our knowledge of the deep sea can proceed, the new race to mine the seafloor is a chance for science, government, and society to avoid making the mistakes that led to the Deepwater Horizon disaster. Erik Cordes told us that, at the very least, the Deepwater Horizon disaster taught a lot of people that there was more to the deep sea than a barren mud plain. "People saw that there was something to explore for," Cordes said. "That was the first major single-point environmental disaster in the deep sea. It was very visible and very public. By the time we found those impacts to the corals I don't know how many people were paying attention—but enough that some of those pictures got out there and people saw them. I think people started to realize, 'Oh, this environmental impact stuff people do in shallow water should be done in the deep sea.'"

In the mid-1990s, world governments agreed to a treaty establishing the International Seabed Authority, which would regulate mining on the seafloor outside the traditional 200-mile "exclusive economic zone" boundaries of most countries. (The United States has not ratified the treaty.)[110] By 2022, more than twenty contractors had permits to explore for minerals in different areas of the international deep sea. Poland, Russia, and France have permits to explore the mid-Atlantic Ridge. Germany, India, Korea, and China are exploring the Indian Ocean. China, Korea, Russia, and Japan are investigating the Northwest Pacific area southeast of Japan. Several European countries, as well as Russia, China, and Japan, are exploring the Clarion-Clipperton Fracture Zone (CCFZ), a deepwater ridge in the mid-Pacific southeast of Hawai'i.

The CCFZ covers over 2 million square miles of seafloor in the northeastern equatorial Pacific. It is considered one of the world's most valuable sites for the potential recovery of metallic crusts and nodules from the seafloor, which include copper, nickel, and cobalt.[111] The *Nautilus* team mapped the CCFZ in high resolution in 2018 in an attempt to fill in a paucity of seafloor image data for this region.[112] In an editorial for a special issue of the journal *Marine Biodiversity*, scientists highlight the biodiversity in the region. Among the novel and barely studied taxa that the scientists document are cirrate (finned) octopuses; long-lived black corals; Xenophyophores, single-celled organisms that accumulate and produce a shell using locally scavenged sand grains; and a diverse array of dinoflagellates, an important component of marine plankton.[113] A recent study of the organisms living in the sediments of the CCFZ documented an abundance of nematodes, copepods, and polychaetes, all feeding primarily on the slow, downward drift of marine snow.[114]

Mining these areas would involve not just the scraping of nodules and deposits off the seafloor but the disturbance and suspension of massive amounts of sedimentary material into the water column, potentially affecting organisms across a wide swath of the mining area.[115] "We've barely explored the darkest realm of the ocean," read the headline of a 2021 *New Yorker* article. "With rare-metal mining on the rise, we're already destroying it."[116] In what now seems like a prescient statement about the threat of industrial colonization of the Pacific, Epeli Hau'ofa wrote in 1994 that "the importance for our ocean for the stability of the global environment, for meeting a significant proportion of the world's protein requirements, for the production of certain marine resources in waters that are relatively clear of pollution, for the global reserves of mineral resources, among others, has been increasingly recognized."[117]

In 2022, Pacific Elders Voice, a group of leaders from Pacific islands, issued an online statement that they were "concerned about a lack of scientific and technical information, in particular, a lack of regulatory capacity on processes for deep seabed mining, and its negative environmental and social impacts to adequately manage and control this type of mining. It is a fact that most of the Pacific Island countries contemplating deep seabed mining do not have the capacity to effectively monitor such mining operations in which they may be involved."[118]

Scientists who specialize in the deep sea are being asked to weigh questions about seafloor exploitation with incomplete knowledge. If there is any lesson we have learned, it's that limiting exploitation on the basis of incomplete information may save far more than we can imagine.[119] In a study published in 2022, an international group of thirty-one scientists called on policymakers to slow down on deep-sea mining. "There's no way we'll have the critical scientific information we need to prevent serious harm and ensure effective protection of the deep ocean in the next two years, or even the next 10 years," said Diva Amon, the lead author, in an article about the study.[120] Amon and her collaborators reviewed 306 peer-reviewed articles on the subject of deep-sea mining to report on what we know and what we don't about deep-sea environments and the effects that mining has on them.[121] They interviewed forty-two people from around the world, including scientists, industry contractors, lawyers, diplomats, and policymakers. Each was asked for their own assessment of the state of scientific knowledge and what more we need to know to mine the deep sea responsibly. The scientists then displayed the results in the form of a 140-cell table, with seven regions of the ocean across the top and twenty areas of scientific knowledge along the side. For each cell, they assigned a color from one of four options, ranging from "scientific knowledge enables evidence-based management" to "there is no or next to no scientific knowledge to enable evidence-based management." It is no surprise that the latter dominates. We have no or next to no scientific knowledge for most of the cells in the table. There is a total lack of information, for every region where there's interest in mining, relating to impacts like contaminant release and toxicity, noise, vibration, light, or cumulative impacts. There is no knowledge for any of the regions that establishes survey and monitoring criteria. We have limited knowledge on what lives in the deep sea, from life histories to where things live, how long they live, and what they eat. Only two cells out of 140 met the criteria for "enables evidence-based management." One is high-resolution bathymetry, which we have of the mid-Atlantic Ridge. The other, for the CCFZ, is knowledge about the impacts of "removal of resources."

Over the last decade, the phrase "we know enough to act" has become something of a mantra among climate scientists and conservation advocates. Do we know everything that it is possible to know about the world and our place in it? Obviously not. Do we know enough to know we need to change our behavior? Obviously we do. We have spent hundreds of thousands of years constructing our mental map of the tides and rocky shorelines and everything that lives in them. We have dived in the coral reefs, watched and understood the finest-scale relationships between the things living in and around them, even destroyed corals and learned to regrow them in labs. We have followed tangled kelp from the earliest days of our venturing out to sea, watched the great undersea forests wither worldwide, and used our knowledge to replant them. We have learned and relearned how to garden for clams and fish. We have pulled fish from the open ocean for tens of thousands of years; realized, at often great cost, that fish were not inexhaustible; and in some places built the protections to bring them back. We have voyaged thousands of miles across the open sea, using our knowledge of wind, waves, stars, currents, and biology to turn seemingly featureless blue horizons into vital and meaningful connective paths. We have mapped the shifting currents of the world's oceans and witnessed the accumulation of our land-based trash in their swirling gyres. We have learned to live on the ice, to thrive in an unforgiving polar environment, and to use our tools to measure with precision as the ice melts and the seas rise. We have invented robots and submarines to take us into the deep sea, to turn it from the lifeless realm of imagination into the world we see of beauty and staggering complexity.

We do not know everything, but we do share an ocean knowledge that is as wide and deep as the seas themselves. This connection is our inheritance as people, a gift from the past to help us navigate the future. It is, all of it, a guide to how to act now.

EPILOGUE

> *I do believe that there is hope for us, and you are very welcome to take some and keep it for your own.*
>
> —Hope Jahren, The Story of More

The story of the white abalone matches the story of many other marine species. A kind of large sea snail, the white abalone once lived from the Baja California peninsula in Mexico to Point Conception, where Southern California takes its sharp turn westward into the Pacific. Abalone connoisseurs widely consider the white abalone the most delicious of the seven abalone species in California, and because it lives on the sand and rocks in 50 to 100 feet of water, preindustrial-era harvest was generally sustainable. Scientists estimate the white abalone's pre-fishing population numbered somewhere between hundreds of thousands to several million.[1]

Then settlers arrived, particularly in the post–World War II era as Southern California boomed, bringing with them the idea of an inexhaustible ocean. New technology allowed the commercial abalone fishery to accelerate in the absence of what turned out to be critical scientific information about how white abalone live. Over the space of just more than a decade, the white abalone that once cobbled the seafloor disappeared. The fishery collapsed by 1980 and the state officially closed it, far too late, in 1996.

Once the fishery was closed, researchers looked for what was left. A submarine survey in 1996 covered nineteen acres of the seafloor over twenty-four dives; the scientists saw nine white abalone in total. A second and larger survey three years later, of once-popular fishing ground, spotted 157. Scientists estimated there might be 2,000 white abalone left alive, in scattered groups of around 100 individuals. Because abalone are broadcast spawners, like coral, they were likely too spread out to reproduce. By the time they were listed as endangered in 2001, they had lost the ability to recover on their own. White abalone can live for perhaps thirty-five to forty years, so the remaining individuals were consigned to live out their long lives in a kind of retirement home. When the U.S. Fish and Wildlife Service formally declared the species endangered, the agency did not declare a corresponding "critical habitat"—a step it is usually required to take legally—for fear any kind of map would serve as an aid to poachers to pick the last few.[2]

Over the next two decades, divers kept watching the remaining abalone as they slowly died off. To save the species from extinction, there was only one option left. In 2000, scientists collected eighteen white abalone to bring back to the lab to try to raise them there, for an eventual return to the wild. It was a more daunting task than perhaps it sounds at first. No one knew how to keep a white abalone healthy in a lab or what to even feed it, let alone how to get one to spawn. If one did spawn, no one knew what to do with the babies. If they could capture them, raise them, feed them, breed them, and then raise the babies . . . no one knew whether it would even work to put those babies back in the ocean.

In writing this book, we realized that we were not simply telling the story of a changing ocean and of the people watching that change; we were in fact telling the story of how our relationship with the ocean would remodel, reshape, and renew. We are entering a new era defined by a paradox: while individual people may have less of a connection to the ocean than ever, our collective human knowledge, fueled by millennia of exploration and centuries of scientific investigation, is greater than it has ever been. Like cartographers slowly filling in the blanks on the map of the world, we have expanded our understanding—of the shape of the seafloor, the physical processes of the water, the biology of creatures in the sun-drenched tide pools, and the freezing lightless depths. And yet right now at conferences, in academic papers, and on social media, an idea has begun to emerge from the holders of this knowledge that somewhere in the last decade, the map of our knowledge about the ocean shifted. The ocean has changed so fast in the last few years, faster than at any point in the entirety of human life on Earth, that our ability to observe struggles to keep pace. Marine scientists now sometimes say that the ocean itself is approaching some new, uncertain threshold: as if it has fallen off the edge of the map.

And yet we have observed examples of people reconnecting themselves and their communities to a future ocean. The problems we face today do not in fact represent a static condition of the ocean but rather (we hope) a turning point. Nothing illustrates this more profoundly than being witness to a new type of science that integrates community, generations of wisdom, hope for the future, and resiliency. We were lucky to witness this kind of science in 2019, as the white abalone recovery program prepared to release young endangered white abalone back into their native environment.

More than twenty years have passed since divers collected the first white abalone from the wild. In 2001, the first generation of rescued abalone spawned, and scientists started to raise the babies. Then a vicious abalone-killing disease, called withering foot syndrome, rampaged through the lab population, killing almost all of them. By 2003, only one of the originally collected abalones was still alive. The scientists running the recovery program split the few dozen juvenile survivors between four labs, to guard against a sudden event wiping out the rest of the population. Ten more years passed, while scientists struggled to keep the abalone alive and get them to spawn. Finally, in 2012, the abalone—and their human caretakers—turned a corner. Kristin Aquilino, who at the time was the lead scientist for the white abalone recovery program, captured some sperm from a male in a lab at the University of California, Santa Barbara. A pairing with eggs from several more abalone yielded 300,000 embryos. Aquilino packed the embryos in Tupperware and an ice chest and drove them through the night to a new captive-rearing facility at the University of California Davis Bodega Marine Laboratory, built specifically for raising abalone.[3] Twenty-four of the embryos survived the year. From those new recruits, over the next few years, the adult population slowly grew. In 2018, Aquilino and her team received permission to capture another nine wild abalone. In spring 2019, the wild-caught abalone and their lab-raised cohort had a massive spawn. For the first time, Aquilino struggled to find a place for all the young abalone. And a few months later, on an otherwise unremarkable day—a typical foggy summer day in Bodega Bay, in a scientific workweek filled with the usual classes, meetings, and research activities at the marine lab—a small group gathered to watch a remarkable event, as 3,000 young white abalone were prepared for a journey back to the ocean.

Aquilino had gathered a small group of people—scientists, conservationists, government officials, Indigenous knowledge holders, and supporters of this program—for a farewell ceremony. The crowd gathered in a makeshift loading dock area, with the door to the white abalone culture laboratory left open to show the view of the light-controlled tanks where the abalone had spent their formative early lives. In the background, open tanks bubbled with water teeming with purple urchins, in preparation for a separate, ongoing experiment. An SUV and a van, back doors standing open and cargo space stacked full with coolers, waited to transport the abalone. After a seven-hour drive to Southern California, divers would swim the abalone to a new home off the coast of Los Angeles.

Aquilino began the small event by thanking all the people who had been involved and the supporters of the program for many years. She read from a blessing she wrote for the departing snails, which included describing the endless work of caring for the snails, all the people who had visited and doted upon them over the years, and a statement of gratitude and hope for their future: "As we asked you questions about how to save your species from our species, you gifted us with curiosity and wonder, with responsibility and purpose, and with hope and optimism."

Jacquelyn Ross joined the ceremony to talk about traditional connections to abalone from the perspective of Pomo and Coast Miwok Indigenous people. In her Northern California homeland, Ross grew up with red abalone. She recalled family members dangling her in the water as a child to watch the abalone stacked up on top of each other. She described searching through the seaweed piles on the beach for the exquisite, miniature shells of young abalone. She worries now that these traditions will disappear with the abalone. "I have cousins in their thirties who have never seen or touched a live abalone in the water; do not know how they like to live; and do not know our fishing ways," Ross wrote in the magazine *Langscape* in 2020. "Another coastal tribe who have recently acquired a gorgeous section of their traditional territory are carefully planning an abalone aquaculture operation to support a research, education, and restoration center. This will be directed toward assisting the red abalone population and other ocean species in regaining strength and vitality. The planning team includes abalone scientists, aquaculture specialists, and engineers for this delicate work. Can it be done? We look to the work underway with the white abalone."[4]

Suki Waters (Kashia Pomo) shared a white abalone prayer that she had written many years earlier, on the morning of the first abalone spawning event for this program. Coast Miwok and Pomo Tribe members had gathered to say goodbye to the abalone who had lived on this land for the beginning of their lives. L. Frank Marquinez then stepped forward to provide a welcome to the abalone, which were being driven to Southern California, the homeland of her Tongva-Acjachmemen tribe. The Tongva-Acjachmemen are not recognized by the federal government. She sang a song in her language. In the small crowd of people standing there, bearing witness to the blessing and the welcoming of the abalone, many closed their eyes to focus on this moment. "Like the abalone, we are also fighting extinction," said Marquinez. "We understand the loneliness of the comeback. But there is also strength in that comeback."

The white abalone are a success story in a world where these stories seem harder and harder to find. Some might simply attribute that success to science and hard work. It was clear that day, standing with scientists, marine lab facilities staff members, local community members, members of multiple different Indigenous groups, and federal agency representatives, that this was a success built on a new kind of future for the ocean. One that builds on the best available knowledge to encourage resilience, and cares for the communities of people whose future is deeply entwined with the survival of every species on this planet. That day was one where we can see a future beyond all the challenges we are presented with today.

As the abalone retuned to the ocean, Aquilino read a poem, which ended with these lines:

> You are the future of your species.
> You are the future of our species.
> Farewell, white abalone.
> This little blue planet is a better place with you in it.

ACKNOWLEDGMENTS

Researching and writing this book through the COVID-19 global pandemic demanded an extraordinary amount from the many people we spoke with. We thank every person who was willing to meet with us in-person as well as everyone who emailed us and shared their piece of the ocean virtually as we researched the book.

We thank our agent Diana Finch for her guidance throughout the book writing and our Columbia University Press editor, Miranda Martin, for her support and an unusual amount of patience with our process. We thank Kaitlyn Kraybill-Voth for beautiful illustrations that brought our stories to life on the page.

We thank S. Whiteaker, G. Grisby, and C. Rocheleau for their research and reference assistance. We appreciate the students in UC Davis GEL/ESP 150C Summer 2022, especially M. Wilmot, L. Chen, and S. Torgerson, for their comments, discussions, and suggestions on the book. We appreciate L. Gaylord, B. Gaylord, and S. McAfee for editing and feedback. We acknowledge the helpful and constructive reviews of two anonymous peer reviewers in improving the manuscript.

We thank Victoria Schlesinger and *Bay Nature* magazine for support in the early stages of research and for publishing pieces that became our chapter on tide pools.

We acknowledge Lia Keener for thoughtful and careful fact-checking of the manuscript.

NOTES

1. THE TIDE POOL

1. John Steinbeck, *The Log from the Sea of Cortez* (London: Penguin, 1951), 49.
2. Jane Lubchenco, "Plant Species Diversity in a Marine Intertidal Community: Importance of Herbivore Food Preference and Algal Competitive Abilities," *American Naturalist* 112, no. 983 (1978): 23–39, https://doi.org/10.1086/283250; Bruce A. Menge and Jane Lubchenco, "On the Genesis of *Community Development and Persistence in a Low Rocky Intertidal Zone*, by Jane Lubchenco and Bruce A. Menge," *Bulletin of the Ecological Society of America* 82, no. 2 (2001): 124–125, https://doi.org/10.2307/20168534.
3. Robert T. Paine, "The Pisaster-Tegula Interaction: Prey Patches, Predator Food Preference, and Intertidal Community Structure," *Ecology* 50, no. 6 (1969): 950–961, https://doi.org/10.2307/1936888; Robert T. Paine, "Food Web Complexity and Species Diversity," *American Naturalist* 100, no. 910 (1966): 65–75, https://doi.org/10.1086/282400.
4. Curtis W. Marean, "When the Sea Saved Humanity," *Scientific American* 303, no. 2 (August 2010): 54–61, http://www.jstor.org/stable/26002131.
5. Jessica E. Tierney, Peter B. deMenocal, and Paul D. Zander, "A Climatic Context for the Out-of-Africa Migration," *Geology* 45, no. 11 (2017): 1023–1026, https://doi.org/10.1130/G39457.1.
6. Curtis W. Marean et al., "Early Human Use of Marine Resources and Pigment in South Africa During the Middle Pleistocene," *Nature* 449, no. 7164 (2007): 905–908, https://doi.org/10.1038/nature06204.
7. Curtis W. Marean, "When the Sea Saved Humanity," *Scientific American* 303, no. 2 (August 2010): 54–61, http://www.jstor.org/stable/26002131.
8. Mark W. Denny and Steven D. Gaines, eds., *Encyclopedia of Tidepools and Rocky Shores* (Berkeley: University of California Press, 2007).
9. Raphael D. Sagarin et al., "Climate-Related Change in an Intertidal Community Over Short and Long Time Scales," *Ecological Monographs* 69, no. 4 (1999): 465–490, https://doi.org/10.1890/0012-9615(1999)069[0465:CRCIAI]2.0.CO;2.
10. Philip Blair Laverty, "Recognizing Indians: Place, Identity, History, and the Federal Acknowledgement of the Ohlone/Costanoan-Esselen Nation" (PhD diss., University of New Mexico, 2010).

11. Sebastián Vizcaíno, *Diary of Sebastián Vizcaíno, 1602–1603*, trans. and ed. Herbert Eugene Bolton (New York: Scribner's, 1916), 91–95.
12. Anderson M. Kat, *Tending the Wild: Native American Knowledge and the Management of California's Natural Resources* (Berkeley: University of California Press, 2013), 98, 176–177.
13. Tsim D. Schneider, *The Archeology of Refuge and Recourse* (Tucson: University of Arizona Press, 2021), 23.
14. Nancy Bartosek, "Conversations with a Tide Pool," *Texas Christian University Magazine*, Summer 2002, https://magazine.tcu.edu/summer-2002/conversations-tide-pool/.
15. Edward F. Ricketts, *Breaking Through: Essays, Journals, and Travelogues of Edward F. Ricketts* (Berkeley: University of California Press, 2006), 13–14.
16. Willis G. Hewatt, "Ecological Studies on Selected Marine Intertidal Communities of Monterey Bay, California," *American Midland Naturalist* 18, no. 2 (1937): 161–206, https://doi.org/10.2307/2420496.
17. James P. Barry et al., "Climate-Related, Long-Term Faunal Changes in a California Rocky Intertidal Community," *Science* 267, no. 5198 (1995): 672–675, https://doi.org/10.1126/science.267.5198.672.
18. Deborah A. Miranda, "'They Were Tough, Those Old Women Before Us': The Power of Gossip in Isabel Meadows's Narratives," *Biography* 39, no. 3 (Summer 2016): 373–401, https://doi.org/10.1353/bio.2016.0047.
19. Linda Yamane, "Lost and Found: Ohlone Culture Comes Home," *Third Force*, August 30, 1995, 18.
20. Yamane, "Lost and Found."
21. "Boats Connect to Monterey Fishing Heritage," Monterey Bay Aquarium, August 4, 2021, https://www.montereybayaquarium.org/stories/boats-connect-monterey-history; David Schmalz, "A Seaside Resident Could Find Little Trace of Her Ancestral Culture, So She Set Out to Re-Create It by Hand," *Monterey County Weekly*, November 1, 2018, https://www.montereycountyweekly.com/people/831/a-seaside-resident-could-find-little-trace-of-her-ancestral-culture-so-she-set-out/article_9def74b4-dd63-11e8-b2d2-37d8fb7c5e3c.amp.html.
22. Throughout this book, we follow the convention set by Jessica Hernandez in her 2022 book, *Fresh Banana Leaves: Healing Indigenous Landscapes Through Indigenous Science*, in that we consider this knowledge and refer to it as *Indigenous science*. While other authors (and even some quoted in this book) might name this *Indigenous knowledge* or *traditional ecological knowledge* (TEK), we follow the guidance of Hernandez on the use of these terms: "In order to heal our environments, which are all Indigenous lands, we must incorporate Indigenous voices, perspectives, and lived experiences. In this book I also refer to these as Indigenous science because they embody our ways of knowing that are rooted from ancestral knowledge and valid sciences. I personally do not like to use traditional ecological knowledge, even though sometimes this is the only way we can refer to our Indigenous knowledge systems, because to me, traditional ecological knowledge places us within past contexts. It is important to note that the same way our environments have adapted, our Indigenous knowledge systems have adapted, and this is why I view it as a science itself—Indigenous science." Jessica Hernandez, *Fresh Banana Leaves: Healing Indigenous Landscapes Through Indigenous Science* (Berkeley, CA: North Atlantic Books, 2022), 13.

1. THE TIDE POOL 219

23. Alejandro Bortolus, James T. Carlton, and Evangelina Schwindt, "Reimagining South American Coasts: Unveiling the Hidden Invasion History of an Iconic Ecological Engineer," *Biodiversity Research* 21, no. 11 (2015): 1267–1283, https://doi.org/10.1111/ddi.12377.
24. Cascade J. B. Sorte et al., "Long-Term Declines in an Intertidal Foundation Species Parallel Shifts in Community Composition," *Global Change Biology* 23, no. 1 (2016): 341–352, https://doi.org/10.1111/gcb.13425.
25. This term was first coined in Ian McHarg, *Design with Nature* (Garden City, NY: Natural History Press), 67–70. See also Daniel Pauly, "Anecdotes and the Shifting Baseline Syndrome of Fisheries," *Trends in Ecology and Evolution* 10, no. 10 (1995): 430.
26. "Snapshot Cal Coast," California Academy of Sciences, www.calacademy.org/calcoast.
27. Sagarin et al., "Climate-Related Change in an Intertidal Community," 465–490.
28. Sarah E. Gilman et al., "A Framework for Community Interactions Under Climate Change," *Trends in Ecology & Evolution* 25, no. 6 (2010): 325–331, https://doi.org/10.1016/j.tree.2010.03.002.
29. Melissa Miner et al., "Large-Scale Impacts of Sea Star Wasting Disease (SSWD) on Intertidal Sea Stars and Implications for Recovery," *PLoS One* 13, no. 3 (2018): e0192870, https://doi.org/10.1371/journal.pone.0192870.
30. Tony Reichhardt, "Clinton's Ocean Agenda Offers Modest Treasures for Science," *Nature* 393, no. 6686 (1998): 609, https://doi.org/10.1038/31287.
31. Phil McKenna, "Science and Adventure in the Spirit of Steinbeck and Ricketts: A New Log from the Sea of Cortez," *Monterey County Weekly*, May 13, 2004, https://www.montereycountyweekly.com/news/local_news/science-and-adventure-in-the-spirit-of-steinbeck-and-ricketts/article_066eb1e6-7140-5ff6-8c0f-90e8330e9561.html.
32. Daniel Stolte, "Remembering Rafe: From Tide Pools to the Stars," *University of Arizona News*, June 3, 2015, https://news.arizona.edu/story/remembering-rafe-from-tide-pools-to-the-stars. The University of Arizona announcement of his death included, "Rafe encouraged all of us to look from the tide pool to the stars and back, to consider the big picture and the small picture, and how they are connected, when trying to make sense of the world. He was fascinated and delighted by what he observed in the tide pools, which, in the words of Steinbeck and Ricketts, were 'ferocious with life: an exuberant fierceness in the littoral here, a vital competition for existence.... They fight back at the sea with a joyful survival.'"
33. Fiorenza Micheli et al., "Field Stations as Sentinels of Change," *Frontiers in Ecology and the Environment* 18, no. 6 (August 2020): 320–322, https://www.jstor.org/stable/26986260.
34. Amanda E. Bates et al., "Biologists Ignore Ocean Weather at Their Peril," *Nature* 560 (2018): 299–301, https://www.nature.com/articles/d41586-018-05869-5.
35. Yamane, "Lost and Found."
36. Emanuele Di Lorenzo and Nathan Mantua, "Multi-Year Persistence of the 2014/15 North Pacific Marine Heatwave," *Nature Climate Change* 6, no. 11 (2016): 1042–1047, https://doi.org/10.1038/nclimate3082.
37. Jacqueline Sones, *The Natural History of Bodega Head* (blog), https://www.bodegahead.blogspot.com/.
38. Eric Sanford et al., "Widespread Shifts in the Coastal Biota of Northern California During the 2014–2016 Marine Heatwaves," *Scientific Reports* 9, no. 1 (2019): 1–14, https://doi.org/10.1038/s41598-019-40784-3.

1. THE TIDE POOL

39. Erica Nielsen and Sam Walkes, "Five Years After Largest Marine Heatwave on Record Hit Northern California Coast, Many Warm-Water Species Have Stuck Around," *Conversation*, October 4, 2021, https://www.theconversation.com/five-years-after-largest-marine-heatwave-on-record-hit-northern-california-coast-many-warm-water-species-have-stuck-around-168053.
40. James T. Fumo et al., "Contextualizing Marine Heatwaves in the Southern California Bight Under Anthropogenic Climate Change," *Journal of Geophysical Research: Oceans* 125, no. 5 (2020): https://doi.org/10.1029/2019JC015674.
41. Rachel Carson, *The Edge of the Sea* (Boston: Houghton Mifflin Harcourt, 1955), 21.

2. THE REEF

1. Jorge Cortés et al., "The CARICOMP Network of Caribbean Marine Laboratories (1985–2007): History, Key Findings, and Lessons Learned," *Frontiers in Marine Science* 5 (2019): 519, https://doi.org/10.3389/fmars.2018.00519.
2. Ivan Nagelkerken et al., "Widespread Disease in Caribbean Sea Fans: I. Spreading and General Characteristics," *Proceedings of the 8th International Coral Reef Symposium*, no. 1 (January 1997): 679–682, https://www.researchgate.net/publication/270285871_Widespread_disease_in_Caribbean_sea_fans_I_Spreading_and_general_characteristics.
3. C. Drew Harvell, "Partial Predation, Inducible Defenses, and the Population Biology of a Marine Bryozoan (Membranipora membranacea)" (PhD diss., University of Washington, 1985).
4. C. Drew Harvell, *Ocean Outbreak* (Oakland: University of California Press, 2021).
5. Garriet W. Smith et al., "Caribbean Sea-Fan Mortalities," *Nature* 383 (1996): 487, https://doi.org/10.1038/383487a0.
6. David M. Geiser et al., "Cause of Sea Fan Death in the West Indies," *Nature* 394, no. 6689 (1998): 137–138, https://doi.org/10.1038/28079.
7. Alisa P. Alker, Garriet W. Smith, and Kiho Kim, "Characterization of Aspergillus sydowii (Thom et Church), a Fungal Pathogen of Caribbean Sea Fan Corals," *Hydrobiologia* 460 (2001): 105–111, https://doi.org/10.1023/A:1013145524136.
8. Harvell, *Ocean Outbreak*, 27.
9. Kiho Kim and C. Drew Harvell, "The Rise and Fall of a Six-Year Coral-Fungal Epizootic," *American Naturalist* 164, no. 5 (November 2004): S52–S63, https://doi.org/10.1086/424609.
10. Harvell, *Ocean Outbreak*, 33.
11. Katie L. Cramer et al., "Widespread Loss of Caribbean Acroporid Corals Was Underway Before Coral Bleaching and Disease Outbreaks," *Science Advances* 6, no. 17 (2020): eaax9395, https://www.publish.csiro.au/mf/MF99078.
12. John Veron et al., "Delineating the Coral Triangle," *Galaxea, Journal of Coral Reef Studies* 11, no. 2 (2009): 91–100, https://www.jstage.jst.go.jp/article/galaxea/11/2/11_2_91/_article.
13. Kelsey L. Sanborn et al., "A New Model of Holocene Reef Initiation and Growth in Response to Sea-Level Rise on the Southern Great Barrier Reef," *Sedimentary Geology* 397 (March 2020): 105556, https://doi.org/10.1016/j.sedgeo.2019.105556.
14. Katarzyna Frankowiak et al., "Photosymbiosis and the Expansion of Shallow-Water Corals," *Science Advances* 2, no. 11 (2016): e1601122, https://doi.org/10.1126/sciadv.1601122.

15. Christopher B. Wall et al., "Divergent Symbiont Communities Determine the Physiology and Nutrition of a Reef Coral Across a Light-Availability Gradient," *International Society for Microbial Ecology Journal* 14, no. 4 (2020): 945–958, https://doi.org/10.1038/s41396-019-0570-1.
16. Ove Hoegh-Guldberg, "The Impact of Climate Change on Coral Reef Ecosystems," *Coral Reefs: An Ecosystem in Transition* (2011): 391–403, https://doi.org/10.1007/978-94-007-0114-4_22.
17. Marjorie L. Reaka-Kudla, "The Global Biodiversity of Coral Reefs: A Comparison with Rainforests," in *Biodiversity II: Understanding and Protecting Our Biological Resources*, ed. Don E. Wilson and Edward O. Wilson (Washington, DC: Joseph Henry Press, 1997), 83–108.
18. Cramer et al., "Widespread Loss of Caribbean Acroporid Corals," eaax9395.
19. Helen E. Fox and Roy L. Caldwell, "Recovery from Blast Fishing on Coral Reefs: A Tale of Two Scales," *Ecological Applications* 16, no. 5 (October 2006): 1631–1635, https://doi.org/10.1890/1051-0761(2006)016[1631:RFBFOC]2.0.CO;2.
20. C. Drew Harvell, *A Sea of Glass* (Oakland: University of California Press, 2019), 30.
21. Peter W. Glynn, "Extensive 'Bleaching' and Death of Reef Corals on the Pacific Coast of Panamá," *Environmental Conservation* 10, no. 2 (Summer 1983): 149–154, https://doi.org/10.1017/S0376892900012248.
22. Ludwig Franzisket, "The Atrophy of Hermatypic Reef Corals Maintained in Darkness and Their Subsequent Regeneration in Light," *International Review of Hydrobiology* 55, no. 1 (1970): 1–12, https://doi.org/10.1002/iroh.19700550102; Ove Hoegh-Guldberg and G. Jason Smith, "The Effect of Sudden Changes in Temperature, Light and Salinity on the Population Density and Export of Zooxanthellae from the Reef Corals Stylophora pistillata Esper and Seriatopora hystrix Dana," *Journal of Experimental Marine Biology and Ecology* 129, no. 3 (August 1989): 279–303, https://doi.org/10.1016/0022-0981(89)90109-3.
23. Paul L. Jokiel and Stephen L. Coles, "Effects of Temperature on the Mortality and Growth of Hawaiian Reef Corals," *Marine Biology* 43 (1977): 201–208, https://doi.org/10.1007/BF00402312.
24. Paul L. Jokiel and Stephen L. Coles, "Effects of Heated Effluent on Hermatypic Corals at Kahe Point, Oahu," *Pacific Science* 28, no. 1 (1974): 1–18, http://hdl.handle.net/10125/1144.
25. Amy McDermott, "A Microscopic Mystery at the Heart of Mass Coral Bleaching," *Proceedings of the National Academy of Sciences* 117, no. 5 (2020): 2232–2235, https://doi.org/10.1073/pnas.1921846117.
26. Ana N. Campoy et al., "The Origin and Correlated Evolution of Symbiosis and Coloniality in Scleractinian Corals," *Frontiers in Marine Science* 7 (2020): 461, https://doi.org/10.3389/fmars.2020.00461.
27. Campoy et al., "The Origin and Correlated Evolution," 461.
28. Terrence J. Blackburn et al., "Zircon U-Pb Geochronology Links the End-Triassic Extinction with the Central Atlantic Magmatic Province," *Science* 340, no. 6135 (2013): 941–945, https://doi.org/10.1126/science.1234204.
29. Alexander M. Dunhill et al., "Impact of the Late Triassic Mass Extinction on Functional Diversity and Composition of Marine Ecosystems," *Paleontology* 61, no. 1 (January 2018): 133–148, https://doi.org/10.1111/pala.12332.

2. THE REEF

30. George Stanley and B. van de Schootbrugge, "The Evolution of the Coral–Algal Symbiosis and Coral Bleaching in the Geologic Past," *Coral Bleaching, Ecological Studies* 233 (2018): 9–26, https://doi.org/10.1007/978-3-319-75393-5_2.
31. Gal Dishon et al., "A Novel Paleo-Bleaching Proxy Using Boron Isotopes and High-Resolution Laser Ablation to Reconstruct Coral Bleaching Events," *Geosciences* 12, no. 19 (2015): 5677–5687, https://doi.org/10.5194/bg-12-5677-2015; Verena Schoepf et al., "Short-Term Coral Bleaching Is Not Recorded by Skeletal Boron Isotopes," *PLoS One* 9, no. 11 (2014): e112011, https://doi.org/10.1371/journal.pone.0112011; Ke-Fu Yu et al., "U-Series Dating of Dead Porites Corals in the South China Sea: Evidence for Episodic Coral Mortality Over the Past Two Centuries," *Quaternary Geochronology* 1, no. 2 (May 2006): 129–141, https://doi.org/10.1016/j.quageo.2006.06.005.
32. Yu et al., "U-Series Dating of Dead Porites Corals," 129–141.
33. Andrew D. Barton and Kenneth S. Casey, "Climatological Context for Large-Scale Coral Bleaching," *Coral Reefs* 24 (2005): 536–554, https://doi.org/10.1007/s00338-005-0017-1.
34. Ernest H. Williams and Lucy Bunkley-Williams, "The World-Wide Coral Reef Bleaching Cycle and Related Sources of Coral Mortality," *Atoll Research Bulletin* 335, no. 1 (1990): 1–71, https://doi.org/10.5479/si.00775630.335.1.
35. Hoegh-Guldberg and Smith, "The Effect of Sudden Changes," 279–303.
36. Ove Hoegh-Guldberg, "Climate Change, Coral Bleaching and the Future of the World's Coral Reefs," *Marine and Freshwater Research* 50, no. 8 (1999): 839–866, https://doi.org/10.1071/MF99078.
37. Janice M. Lough, "Small Change, Big Difference: Sea Surface Temperature Distributions for Tropical Coral Reef Ecosystems, 1950–2011," *Journal of Geophysical Research* 117, no. C9 (September 2012): C09018, https://doi.org/10.1029/2012JC008199.
38. Irus Braverman, *Coral Whisperers: Scientists on the Brink*, ed. Julie Guthman, Jake Kosek, and Rebecca Lave (Oakland: University of California Press, 2018), 56.
39. "The 1997–98 El Niño," National Oceanic and Atmospheric Administration, last modified October 24, 2022, www.celebrating200years.noaa.gov/magazine/enso/el_nino.html.
40. David B. Enfield and Dennis A. Mayer, "Tropical Atlantic Sea Surface Temperature Variability and Its Relation to El Niño-Southern Oscillation," *Journal of Geophysical Research* 102, no. C1 (January 1997): 929–945, https://doi.org/10.1029/96JC03296.
41. Shannon Sully et al., "A Global Analysis of Coral Bleaching over the Past Two Decades," *Nature Communications* 10, no. 1 (2019): 1264, https://doi.org/10.1038/s41467-019-09238-2.
42. Laura D. Mydlarz et al., "Cellular Responses in Sea Fan Corals: Granular Amoebocytes React to Pathogen and Climate Stressors," *PLoS One* 3, no. 3 (2008): e1811, https://doi.org/10.1371/journal.pone.0001811.
43. C. Drew Harvell et al., "Emerging Marine Diseases—Climate Links and Anthropogenic Factors," *Science* 285, no. 5433 (1999): 1505–1510, https://doi.org/10.1126/science.285.5433.1505.
44. Hoegh-Guldberg, "Climate Change, Coral Bleaching," 839–866.
45. Rebecca Albright, "Scientists Are Taking Extreme Steps to Help Corals Survive," *Scientific American*, January 1, 2018, https://www.scientificamerican.com/article/scientists-are-taking-extreme-steps-to-help-corals-survive/.
46. "Coral Bleaching During & Since the 2014–2017 Global Coral Bleaching Event Status and an Appeal for Observations," National Oceanic and Atmospheric Administration Satellite and Information Service, last modified March 19, 2018, www.coralreefwatch.noaa.gov/satellite/analyses_guidance/global_coral_bleaching_2014-17_status.php.

47. Terry Hughes (@ProfTerryHughes), "I showed the results of aerial surveys of #bleaching on the #GreatBarrierReef to my students, And then we wept," Twitter, April 19, 2016, www.twitter.com/profterryhughes/status/722512223067721728?lang=en.
48. Graham Readfearn, "Great Barrier Reef's Third Mass Bleaching in Five Years the Most Widespread Yet," *Guardian*, April 6, 2020, www.theguardian.com/environment/2020/apr/07/great-barrier-reefs-third-mass-bleaching-in-five-years-the-most-widespread-ever.
49. Lauren Sommer, "Australia's Great Barrier Reef Is Hit with Mass Coral Bleaching Yet Again," *National Public Radio*, March 26, 2022, www.npr.org/2022/03/26/1088886918/australia-great-barrier-reef-coral-bleaching-climate.
50. *Great Barrier Reef Marine Park Act* 1975 (Cth), www.legislation.gov.au/Details/C2011C00149.
51. Nicholas A. J. Graham et al., "Changing Role of Coral Reef Marine Reserves in a Warming Climate," *Nature Communications* 11, no. 1 (2020): 2000, https://doi.org/10.1038/s41467-020-15863-z.
52. Kim Cobb, "Stand up for Science Rally" (presentation, Boston, December 13, 2016), www.youtube.com/watch?v=8A266h2LhNU&t=118s.
53. *Before the House of Representatives Committee on Natural Resources: Climate Change: Impacts and the Need to Act*, 113th Congress, 64 (2014) (Statement of Kim Cobb, climate scientist), www.naturalresources.house.gov/imo/media/doc/Cobb_NaturalResources_feb19_written.pdf.
54. "Trends in Atmospheric Carbon Dioxide," National Oceanic and Atmospheric Administration Global Monitoring Laboratory, accessed April 28, 2023, www.gml.noaa.gov/ccgg/trends/mlo.html.
55. Joan A. Kleypas et al., "Geochemical Consequences of Increased Atmospheric Carbon Dioxide on Coral Reefs," *Science* 284, no. 5411 (1999): 118–120, https://doi.org/10.1126/science.284.5411.118.
56. Joan A. Kleypas, John W. McManus, and Lambert A. B. Meñez, "Environmental Limits to Coral Reef Development: Where Do We Draw the Line?," *American Zoologist* 39, no. 1 (February 1999): 146–159, https://doi.org/10.1093/icb/39.1.146.
57. Kleypas et al., "Geochemical Consequences of Increased Atmospheric Carbon Dioxide," 118–120.
58. A note to the reader that our discussion with Joanie was in 2018, so this is not a reference to a popular movie released in 2022.
59. Ken Caldeira and Michael E. Wickett, "Anthropogenic Carbon and Ocean pH," *Nature* 425, no. 6956 (2003): 365, https://doi.org/10.1038/425365a.
60. Rebecca Albright et al., "Reversal of Ocean Acidification Enhances Net Coral Reef Calcification," *Nature* 531, no. 7594 (2016): 362–365, https://doi.org/10.1038/nature17155.
61. Rebecca Albright et al., "Carbon Dioxide Addition to Coral Reef Waters Suppresses Net Community Calcification," *Nature* 555, no. 7697 (2018): 516–519, https://doi.org/10.1038/nature25968.
62. Irus Braverman, *Coral Whisperers: Scientists on the Brink*, ed. Julie Guthman, Jake Kosek, and Rebecca Lave (Oakland: University of California Press, 2018), 251.
63. Ed Yong, "How Coral Researchers Are Coping with the Death of Reefs," *Atlantic*, November 21, 2017, https://www.theatlantic.com/science/archive/2017/11/coral-scientists-coping-reefs-mental-health/546440/.
64. Oliver Milman, "Earth Has Lost a Third of Arable Land in Past 40 Years, Scientists Say," *Guardian*, December 2, 2015, https://www.theguardian.com/environment/2015/dec/02/arable-land-soil-food-security-shortage.

65. Braverman, *Coral Whisperers*, 238.
66. Paulina Kaniewska et al., "Signaling Cascades and the Importance of Moonlight in Coral Broadcast Mass Spawning," *eLife* 4 (2015): e09991, https://doi.org/10.7554/eLife.09991; Adam M. Reitzel, Ann M. Tarrant, and Oren Levy, "Circadian Clocks in the Cnidaria: Environmental Entrainment, Molecular Regulation, and Organismal Outputs," *Integrative and Comparative Biology* 53, no. 1 (July 2013): 118–130, https://doi.org/10.1093/icb/ict024.
67. Michelle Nijhuis, "Spawning an Intervention," *bioGraphic*, September 26, 2018, www.biographic.com/spawning-an-intervention/.
68. "Watch Our Corals Spawn!," *California Academy of Sciences*, May 15, 2020, www.calacademy.org/watch-our-corals-spawn.
69. Irus Braverman, "The Pristine Is Gone: An Interview with Jeremy Jackson," in *Coral Whisperers: Scientist on the Brink* (Oakland: University of California Press, 2018), 104.
70. Emily S. Darling et al., "Social-Environmental Drivers Inform Strategic Management of Coral Reefs in the Anthropocene," *Nature Ecology & Evolution* 3, no. 9 (2019): 1341–1350, https://doi.org/10.1038/s41559-019-0953-8; E. Sala, "A Pacific Rebirth," *National Geographic*, November 2022, 104–121.
71. Tony Moore, "IVF Coral Spawns for First Time, Paving Way for Great Barrier Reef Repair," *Sydney Morning Herald*, December 6, 2021, https://www.smh.com.au/environment/climate-change/ivf-coral-spawns-for-first-time-paving-way-for-great-barrier-reef-repair-20211202-p59eaa.html.
72. Cramer et al., "Widespread Loss of Caribbean Acroporid Corals," eaax9395.
73. Harilaos A. Lessios, "The Great *Diadema antillarum* Die-Off: 30 Years Later," *Annual Review of Marine Science* 8, no. 1 (2016): 267–283, https://doi.org/10.1146/annurev-marine-122414-033857.
74. Peter J. Edmunds and Robert C. Carpenter, "Recovery of *Diadema antillarum* Reduces Macroalgal Cover and Increases Abundance of Juvenile Corals on a Caribbean Reef," *Proceedings of the National Academy of Sciences* 98, no. 9 (2001): 5067–5071, https://doi.org/10.1073/pnas.071524598.
75. Stacey M. Williams, "The Reduction of Harmful Algae on Caribbean Coral Reefs Through the Reintroduction of a Keystone Herbivore, the Long Spined Sea Urchin, *Diadema antillarum*," *Restoration Ecology* 30, no. 1 (January 2021): e13475, https://doi.org/10.1111/rec.13475.
76. "Diadema Response Network—Map of Diadema and Other Sea Urchins in the Caribbean," Atlantic and Gulf Rapid Reef Assessment, accessed November 25, 2022, www.agrra.org.

3. THE FOREST

1. Scott Bennett et al., "The 'Great Southern Reef': Social, Ecological and Economic Value of Australia's neglected Kelp Forests," *Marine and Freshwater Research* 67, no. 1 (2015): 47–56, https://doi.org/10.1071/MF15232.
2. Sylvia A. Earle, "Undersea World of a Kelp Forest," *National Geographic* 158, no. 3 (1980): 411–426.

3. Brian Gaylord et al., "Spatial Patterns of Flow and Their Modification Within and Around a Giant Kelp Forest," *Limnology and Oceanography* 52, no. 5 (2007): 1838–1852, https://doi.org/10.4319/lo.2007.52.5.1838.
4. Wheeler J. North and Bates Littlehales, "Giant Kelp: Sequoias of the Sea," *National Geographic* 142 (August 1972): 250–269.
5. Darryl Fears, "On Land, Australia's Rising Heat Is 'Apocalyptic.' In the Ocean, It's Worse," *Washington Post*, December 27, 2019, https://www.washingtonpost.com/graphics/2019/world/climate-environment/climate-change-tasmania/.
6. Thomas Wernberg et al., "Status and Trends for the World's Kelp Forests," *World Seas: An Environmental Evaluation* 3 (2019): 57–78, https://doi.org/10.1016/B978-0-12-805052-1.00003-6.
7. John J. Bolton, "The Biogeography of Kelps (Laminariales, Phaeophyceae): A Global Analysis with New Insights from Recent Advances in Molecular Phylogenetics," *Helgoland Marine Research* 64 (2010): 263–279, https://doi.org/10.1007/s10152-010-0211-6.
8. Avery P. Hill et al., "Low-Elevation Conifers in California's Sierra Nevada Are out of Equilibrium with Climate," *Proceedings of the National Academy of Sciences Nexus* 2, no. 2 (2023): 1–9, https://doi.org/10.1093/pnasnexus/pgad004; Zach St. George, *The Journeys of Trees: A Story About Forests, People, and the Future* (New York City: W. W. Norton, 2020), 3–4.
9. Laura Rogers-Bennett and Cynthia A. Catton, "Marine Heat Wave and Multiple Stressors Tip Bull Kelp Forest to Sea Urchin Barrens," *Scientific Reports* 9, no. 1 (2019): 15050, https://doi.org/10.1038/s41598-019-51114-y; Nur Arafeh-Dalmau et al., "Marine Heat Waves Threaten Kelp Forests," *Science* 367, no. 6478 (2020): 635, https://doi.org/10.1126/science.aba5244.
10. Cayne Layton et al., "Kelp Forest Restoration in Australia," *Frontiers in Marine Science* 7 (2020): 74, https://doi.org/10.3389/fmars.2020.00074.
11. Adriana Vergés et al., "Operation Crayweed: Ecological and Sociocultural Aspects of Restoring Sydney's Underwater Forests," *Ecological Management and Restoration* 21, no. 2 (May 2020): 74–85, https://doi.org/10.1111/emr.12413.
12. North and Littlehales, "Giant Kelp."
13. Wheeler J. North and John S. Pearse, "Sea Urchin Population Explosion in Southern California Coastal Waters," *Science* 167, no. 3915 (1970): 209, https://doi.org/10.1126/science.167.3915.209.b.
14. Mia J. Tegner and Paul K. Dayton, "Sea Urchins, El Niños, and the Long-Term Stability of Southern California Kelp Forest Communities," *Marine Ecology Progress Series* 77, no. 1 (1991): 49–63.
15. Anthony Larkum, Michelle Waycott, and John G. Conran, "Evolution and Biogeography of Seagrasses," *Seagrasses of Australia* (2018): 3–29, https://doi.org/10.1007/978-3-319-71354-0_1.
16. Catherine J. Collier, Sven Uthicke, and Michelle Waycott, "Thermal Tolerance of Two Seagrass Species at Contrasting Light Levels: Implications for Future Distribution in the Great Barrier Reef," *Limnology and Oceanography* 56, no. 6 (2011): 2200–2210, https://doi.org/10.4319/lo.2011.56.6.2200.
17. Maarten J. M. Christenhusz and James W. Byng, "The Number of Known Plants Species in the World and Its Annual Increase," *Phytotaxa* 261, no. 3 (2016): 201–217, https://doi.org/10.11646/phytotaxa.261.3.1.

18. U.S. Department of the Interior, "Seagrass Meadows," National Parks Service, https://www.nps.gov/subjects/oceans/seagrass-meadows.htm.
19. Richard K. F. Unsworth et al., "Global Challenges for Seagrass Conservation," *Ambio* 48 (2019): 801–815, https://doi.org/10.1007/s13280-018-1115-y; J. Emmett Duffy et al., "Toward a Coordinated Global Observing System for Seagrasses and Marine Macroalgae," *Frontiers in Marine Science* 6 (2019): 317, https://doi.org/10.3389/fmars.2019.00317.
20. Richard K. F. Unsworth et al., "Seagrass Meadows Support Global Fisheries Production," *Conservation Letters* 12, no. 1 (2019): e12566, https://doi.org/10.1111/conl.12566.
21. Carlos M. Duarte, "The Future of Seagrass Meadows," *Environmental Conservation* 29, no. 2 (2002): 192–206, https://doi.org/10.1017/S0376892902000127; James W. Fourqurean et al., "Seagrass Ecosystems as a Globally Significant Carbon Stock," *Nature Geoscience* 5, no. 7 (2012): 505–509, https://doi.org/10.1038/ngeo1477.
22. Erin Foster et al., "Physical Disturbance by Recovering Sea Otter Populations Increases Eelgrass Genetic Diversity," *Science* 374, no. 6565 (2021): 333–336, https://doi.org/10.1126/science.abf2343; Brent B. Hughes et al., "Recovery of a Top Predator Mediates Negative Eutrophic Effects on Seagrass," *Proceedings of the National Academy of Sciences* 110, no. 38 (2013): 15313–15318, https://doi.org/10.1073/pnas.1302805110.
23. Duarte, "The Future of Seagrass Meadows."
24. Michelle Waycott et al., "Accelerating Loss of Seagrasses Across the Globe Threatens Coastal Ecosystems," *Proceedings of the National Academy of Sciences* 106, no. 30 (2009): 12377–12381, https://doi.org/10.1073/pnas.0905620106.
25. Kira A. Krumhansl et al., "Global Patterns of Kelp Forest Change over the Past Half-Century," *Proceedings of the National Academy of Sciences* 113, no. 16 (2016): 13785–13790, https://doi.org/10.1073/pnas.1606102113.
26. Earle, "Undersea World of a Kelp Forest."
27. Matthew Edwards et al., "Marine Deforestation Leads to Widespread Loss of Ecosystem Function," *Public Library of Science One* 15, no. 3 (2020): e0226173, https://doi.org/10.1371/journal.pone.0226173.
28. Jesse Farmer et al., "The Bering Strait was Flooded 10,000 Years Before the Last Glacial Maximum," *Proceedings of the National Academy of Sciences* 120, no. 1 (2023): e2206742119, https://doi.org/10.1073/pnas.2206742119.
29. James Adovasio and David Pedler, "Monte Verde and the Antiquity of Humankind in the Americas," *Antiquity* 71, no. 273 (1997): 573–580, https://doi.org/10.1017/S0003598X00085331.
30. Tom Dillehay et al., "Monte Verde: Seaweed, Food, Medicine, and the Peopling of South America," *Science* 320, no. 5877 (2008): 784–786, https://doi.org/10.1126/science.1156533.
31. Darcy L. Mathews and Nancy J. Turner, "Ocean Cultures: Northwest Coast Ecosystems and Indigenous Management Systems," in *Conservation for the Anthropocene Ocean: Interdisciplinary Science in Support of Nature and People*, ed. Phillip S. Levin and Melissa R. Poe (Cambridge, MA: Academic Press, 2017), 169–206.
32. Jon M. Erlandson et al., "The Kelp Highway Hypothesis: Marine Ecology, the Coastal Migration Theory, and the Peopling of the Americas," *Journal of Island and Coastal Archaeology* 2, no. 2 (2007): 161–174, https://doi.org/10.1080/15564890701628612.

33. William Peterson, Marie Robert, Nicholas Bond et al., "The Warm Blob—Conditions in the Northeastern Pacific Ocean," *PICES Press* 23, no. 1 (Winter 2015): 36–38.
34. Kisei R. Tanaka and Kyle Van Houtan, "The Recent Normalization of Historical Marine Heat Extremes," *Public Library of Science Climate* 1, no. 2 (2022): e0000007, https://doi.org/10.1371/journal.pclm.0000007.
35. Michael G. Jacox et al., "Forcing of Multiyear Extreme Ocean Temperatures That Impacted California Current Living Marine Resources in 2016," *Bulletin of the American Meteorological Society* 99, no. 1 (January 2018): S27-S33, https://doi.org/10.1175/BAMS-D-17-0119.1.
36. William Peterson, Marie Robert, Nicholas Bond et al., "The Blob (Part Three): Going, Going, Gone?," *PICES Press* 24, no. 1 (Winter 2016): 46–48.
37. William Peterson, Xiuning Du, and Jennifer Fisher, "Effects of the Warm Blob on Phytoplankton and Zooplankton off Central Oregon" (presentation, San Diego, CA, 2016), https://meetings.pices.int/Publications/Presentations/PICES-2016/S12-Peterson.pdf.
38. Monica Garsky, "Sea Lion Pup Found Sleeping in Booth in Marine Room Restaurant," *Nation Broadcasting Corporation: San Diego*, February 5, 2016, https://www.nbcsandiego.com/news/local/seaworld-san-diego-sea-lion-pup-rescue-marine-room-restaurant/63582/.
39. Ryan M. McCabe et al., "An Unprecedented Coastwide Toxic Algal Bloom Linked to Anomalous Ocean Conditions," *Geophysical Research Letters* 43, no. 19 (2016): 10336–10376, https://doi.org/10.1002/2016GL070023.
40. Rogers-Bennett and Catton, "Marine Heat Wave and Multiple Stressors."
41. "SSWS Updates," Multi-Agency Rocky Intertidal Network, October 24, 2013, https://marine.ucsc.edu/data-products/sea-star-wasting/updates.html#Oct24_2013.
42. Ian Hewson et al., "Investigating the Complex Association Between Viral Ecology, Environment, and Northeast Pacific Sea Star Wasting," *Frontiers in Marine Science* 5 (2018): 77, https://doi.org/10.3389/fmars.2018.00077.
43. Bruce A. Menge et al., "Sea Star Wasting Disease in the Keystone Predator *Pisaster ochraceus* in Oregon: Insights into Differential Population Impacts, Recovery, Predation Rate, and Temperature Effects from Long-Term Research," *Public Library of Science One* 11, no. 5 (2016): e0153994, https://doi.org/10.1371/journal.pone.0153994.
44. Citlalli A. Aquino et al., "Evidence That Microorganisms at the Animal-Water Interface Drive Sea Star Wasting Disease," *Frontiers in Microbiology* 11 (2021): 3278, https://doi.org/10.3389/fmicb.2020.610009.
45. Matt Mcknight, "Scientists Grow Sea Stars in Lab to Understand Mass Die-Off Along Pacific Coast," Reuters, February 17, 2023, https://www.reuters.com/business/environment/scientists-grow-sea-stars-lab-understand-mass-die-off-along-pacific-coast-2023-02-16/.
46. C. Drew Harvell et al., "Disease Epidemic and a Marine Heat Wave Are Associated with the Continental-Scale Collapse of a Pivotal Predator (*Pycnopodia helianthoides*)," *Science Advances* 5, no. 1 (2019): eaau7042, https://doi.org/10.1126/sciadv.aau7042.
47. Sierra Garcia, "These Urchin Slayers Are Trying to Save California's Underwater 'Rainforest,'" Grist, September 9, 2021, www.grist.org/climate/zombie-purple-urchin-california-kelp-forest-climate/.
48. Aaron Eger et al., "Global Kelp Forest Restoration: Past Lessons, Status, and Future Goals," *Biological Reviews* (2021): https://doi.org/10.1111/brv.12850.

49. The Bay Foundation and Vantuna Research Group, *Palos Verdes Kelp Forest Restoration Project: Project Year 9: July 2021–June 2022* (Sacramento: California Department of Fish and Wildlife, 2022).
50. Hans K. Strand et al., "Optimizing the Use of Quicklime (CaO) for Sea Urchin Management—A Lab and Field Study," *Ecological Engineering* 143 (2020): 100018, https://doi.org/10.1016/j.ecoena.2020.100018.
51. Taegeon Oh et al., "The Behavioral Response of Purple Sea Urchins *Heliocidaris crassispina* to Food and an Electrical Stimulus," *Korean Journal of Fisheries and Aquatic Sciences* 52, no. 5 (2019): 524–533, https://10.5657/KFAS.2019.0524.
52. Tara Duggan, "Satellite Images Show Kelp Forest Has Doubled in Size on California's North Coast, After a Dramatic Collapse," *San Francisco Chronicle*, November 5, 2021, https://www.sfchronicle.com/climate/article/Satellite-images-show-kelp-forest-has-doubled-in-16589392.php.
53. Jennifer E. Smith et al., "Spiny Lobsters Prefer Native Prey over Range-Extending Invasive Urchins," *International Council for the Exploration of the Sea Journal of Marine Science* 79, no. 4 (May 2022): 1353–1362, https://doi.org/10.1093/icesjms/fsac058.
54. Melinda A. Coleman et al., "Absence of a Large Brown Macroalga on Urbanized Rocky Reefs Around Sydney, Australia, and Evidence for Historical Decline," *Journal of Phycology* 44, no. 4 (2008): 897–901, https://doi.org/10.1111/j.1529-8817.2008.00541.x.
55. Vergés et al., "Operation Crayweed."
56. Vergés et al., "Operation Crayweed."
57. Kuen Y. Lee and David J. Mooney, "Alginate: Properties and Biomedical Applications," *Progress in Polymer Science* 37, no. 1 (2012): 106–126, https://doi.org/10.1016/j.progpolymsci.2011.06.003.
58. National Oceanic and Atmospheric Administration, "How Do People Use Kelp?," National Ocean Service Website, last modified January 20, 2023, www.oceanservice.noaa.gov/facts/pplkelp.html.
59. Julio A. Vázquez, "The Brown Seaweeds Fishery in Chile," *Fisheries and Aquaculture in the Modern World* (2016): 123–141, https://doi.org/10.1016/10.5772/62876.
60. Kira A. Krumhansl et al., "Global Patterns of Kelp Forest Change over the Past Half-Century," *Proceedings of the National Academy of Sciences* 113, no. 48 (2016): 13785–13790, https://doi.org/10.1073/pnas.1606102113; Julio A. Vázquez, "Production, Use and Fate of Chilean Brown Seaweeds: Resources for a Sustainable Fishery" (presentation, Nineteenth International Seaweed Symposium: Proceedings of the 19th International Seaweed Symposium, Kobe, Japan, March 2007), https://doi.org/10.1007/978-1-4020-9619-8_2.
61. Council of the Haida Nation, "History of the Haida Nation," https://www.haidanation.ca/haida-nation/.
62. Barbara Wilson, "DamXan gud.ad t'alang hllGang.gulXads Gina Tllgaay (Working Together to Make It a Better World)" (master's thesis, Simon Fraser University, 2019).
63. Haida Marine Traditional Knowledge Study Participants et al., *Haida Marine Traditional Knowledge Study, Volume Two: Seascape Unit Summary* (British Columbia: Council of the Haida Nation, 2011), https://docslib.org/doc/12161781/haida-marine-traditional-knowledge-study-report-volume-2.
64. Joshua L. Reid, *The Sea Is My Country: The Maritime World of the Makahs, an Indigenous Borderlands People* (New Haven, CT: Yale University Press, 2015), 39–40.

65. Lynn Lee, "Collaborative Marine Monitoring and Coastal Habitat Restoration Within an Indigenous Co-Management Context in Gwaii Haanas, Haida Gwaii, Canada" (presentation, State Estuary & Ocean Science Center, San Francisco, CA, October 2020).
66. Hamdi Issawi, "Sea Otters Are Back with a Worrying Vengeance in B.C.," *Maclean's*, October 15, 2020, https://www.macleans.ca/society/environment/sea-otters-are-back-with-a-worrying-vengeance-in-b-c/.
67. Anne K. Salomon et al., "Coastal Voices," accessed in 2018 and on December 20, 2022, www.coastalvoices.net.
68. Reid, *The Sea Is My Country*, 12.
69. John Kirlin et al., "California's Marine Life Protection Act Initiative: Supporting Implementation of Legislation Establishing a Statewide Network of Marine Protected Areas," *Ocean and Coastal Management* 74 (2013): 3–13, https://doi.org/10.1016/j.ocecoaman.2012.08.015.

4. THE GARDENS

1. Douglas Deur, Kim Recalma-Clutesi, and William White, "Benediction: The Teachings of Chief Kwaxsistalla Adam Dick and the Atla'gimma ('Spirits of the Forest') Dance," in *Plants, People, and Places: The Roles of Ethnobotany and Ethnoecology in Indigenous Peoples' Land Rights in Canada and Beyond*, ed. Nancy J. Turner (Montreal: McGill-Queen's University Press, 2020), xvii–xxv; Kim Recalma-Clutesi and Doug Deur, in discussion with the authors, 2020 and 2022.
2. U'mista Cultural Society, "The Potlatch Ban," Living Tradition: The Kwakwaka'wakw Potlatch on the Northwest Coast, accessed April 26, 2023, https://umistapotlatch.ca/potlatch_interdire-potlatch_ban-eng.php; Kim Recalma-Clutesi and Doug Deur, in discussion with the authors, 2022.
3. Douglas Deur, Kim Recalma-Clutesi, and William White, *Plants, People, and Places: The Roles of Ethnobotany and Ethnoecology in Indigenous Peoples' Land Rights in Canada and Beyond*, ed. Nancy J. Turner (Montreal: McGill-Queen's University Press, 2020), 1–384.
4. "State Listed Species," Washington Department of Fish and Wildlife, last modified April 2023, www.wdfw.wa.gov/species-habitats/at-risk/listed.
5. "Southern Resident Killer Whales," United States Environmental Protection Agency, last modified June 2021, https://www.epa.gov/salish-sea/southern-resident-killer-whales.
6. C. Drew Harvell, *Ocean Outbreak* (Oakland: University of California Press, 2021), 12.
7. Jon M. Erlandson and Torben C. Rick, "Archaeology Meets Marine Ecology: The Antiquity of Maritime Cultures and Human Impacts on Marine Fisheries and Ecosystems," *Annual Review of Marine Science* 2 (2010): 231–251, https://doi.org/10.1146/annurev.marine.010908.163749.
8. Curtis W. Marean et al., "Early Human Use of Marine Resources and Pigment in South Africa During the Middle Pleistocene," *Nature* 449, no. 7164 (2007): 905–908, https://doi.org/10.1038/nature06204.
9. Tsim D. Schneider, *The Archeology of Refuge and Recourse* (Tucson: University of Arizona Press, 2021), 161.
10. Dana Lepofsky et al., "Ancient Shellfish Mariculture on the Northwest Coast of North America," *American Antiquity* 80, no. 2 (2015): 236–259, https://doi.org/10.7183/0002-7316.80.2.236.

11. Junko Habu et al., "Shell Midden Archaeology in Japan: Aquatic Food Acquisition and Long-Term Change in the Jomon Culture," *Quaternary International* 239, no. 1–2 (2011): 19–27, https://doi.org/10.1016/j.quaint.2011.03.014.
12. Robert Günther, "The Oyster Culture of the Ancient Romans," *Journal of the Marine Biological Association of the United Kingdom* 4, no. 4 (1897): 360–365, https://doi.org/10.1017/S0025315400005488.
13. Lepofsky et al., "Ancient Shellfish Mariculture"; Wendy Coy, "Discoveries at Toms Point Alter Timeline for Native Americans in California," *Audubon Canyon Ranch*, September 16, 2019, https://egret.org/discoveries-at-toms-point-alter-timeline-for-native-americans-in-california/.
14. John R. Harper, Janet Haggerty, and Mary C. Morris, *Final Report: Broughton Archipelago Clam Terrace Survey* (Sidney: Coastal and Ocean Resources, 1995).
15. Harper, Haggerty, and Morris, *Final Report: Broughton Archipelago Clam Terrace Survey*.
16. Dana Lepofsky et al., "Ancient Anthropogenic Clam Gardens of the Northwest Coast Expand Clam Habitat," *Ecosystems* 24 (2021): 248–260, https://doi.org/10.1007/s10021-020-00515-6; Nicole F. Smith et al., "3500 Years of Shellfish Mariculture on the Northwest Coast of North America," *PLoS One* 14, no. 2 (2019): e0211194, https://doi.org/10.1371/journal.pone.0211194.
17. Dana Lepofsky et al., *Ancient Shellfish Mariculture on the Pacific Northwest Coast* (Cambridge: Cambridge University Press, 2017), 2. Including this quote: "As several archaeologists have pointed out . . ., the accessibility of many shellfish also makes them vulnerable to overharvesting. Archaeological literature on shellfish is replete with documentation of the effects of overharvesting, evidenced by declining shell abundance and size of targeted/harvested species through time . . ., and resource switching to lower ranked shellfish."
18. Dana Lepofsky and Megan Caldwell, "Indigenous Marine Resource Management on the Northwest Coast of North America," *Ecological Processes* 2, no. 1 (2013): 1–12, https://doi.org/10.1186/2192-1709-2-12.
19. Douglas Deur et al., "Kwakwaka'wakw ('Clam Gardens')," *Human Ecology* 43 (2015): 201–212, https://doi.org/10.1007/s10745-015-9743-3.
20. Amy Groesbeck et al., "Ancient Clam Gardens Increased Shellfish Production: Adaptive Strategies from the Past Can Inform Food Security Today," *PLoS One* 9, no. 3 (2014): e91235, https://doi.org/10.1371/journal.pone.0091235.
21. Darcy L. Mathews and Nancy J. Turner, "Ocean Cultures: Northwest Coast Ecosystems and Indigenous Management Systems," *Conservation for the Anthropocene Ocean: Interdisciplinary Science in Support of Nature and People* (2017): 169–206, https://doi.org/10.1016/B978-0-12-805375-1.00009-X.
22. Jessica Hernandez, *Fresh Banana Leaves: Healing Indigenous Landscapes Through Indigenous Science* (Berkeley, CA: North Atlantic Books, 2022).
23. Erin Slade, Iain McKechnie, and Anne K. Salomon, "Archaeological and Contemporary Evidence Indicates Low Sea Otter Prevalence on the Pacific Northwest Coast During the Late Holocene," *Ecosystems* 25 (2022): 548–566, https://doi.org/10.1007/s10021-021-00671-3.
24. Ginerva Toniello et al., "11,500 y of Human–Clam Relationships Provide Long-Term Context for Intertidal Management in the Salish Sea, British Columbia," *Proceedings of*

the *National Academy of Sciences* 166, no. 44 (2019): 22106–22114, https://doi.org/10.1073/pnas.1905921116.

25. "2012 Symposium on the Ocean in a High-CO_2 World," Ocean Acidification, accessed April 27, 2023, http://ocean-acidification.net/international-symposia/2012-symposium-on-the-ocean-in-a-high-co$_2$-world/.

26. Elizabeth Grossman, "Northwest Oyster Die-Offs Show Ocean Acidification Has Arrived," Yale Environment 360, last modified November 21, 2011, http://www.e360.yale.edu/features/northwest_oyster_die-offs_show_ocean_acidification_has_arrived.

27. Grossman, "Northwest Oyster Die-Offs."

28. Ken Caldeira and Michael E. Wickett, "Anthropogenic Carbon and Ocean pH," *Nature* 425, no. 6956 (2003): 365, https://doi.org/10.1038/425365a; Scott C. Doney et al., "Ocean Acidification: The Other CO_2 Problem," *Annual Review of Marine Science* 1 (2009): 169–192, https://doi.org/10.1146/annurev.marine.010908.163834.

29. Joan A. Kleypas et al., "Geochemical Consequences of Increased Carbon Dioxide on Coral Reefs," *Science* 284, no. 5411 (1999): 118–120, https://doi.org/10.1126/science.284.5411.118.

30. Kristy J. Kroeker et al., "Impacts of Ocean Acidification on Marine Organisms: Quantifying Sensitivities and Interaction with Warming," *Global Change Biology* 19, no. 6 (2013): 1884–1896, https://doi.org/10.1111/gcb.12179.

31. Brittany M. Jellison et al., "Ocean Acidification Alters the Response of Intertidal Snails to a Key Sea Star Predator," *Royal Society Publishing B: Biological Sciences* 283, no. 1833 (2016): 20160890, https://doi.org/10.1098/rspb.2016.0890; Rachael M. Heuer and Martin Grosell, "Physiological Impacts of Elevated Carbon Dioxide and Ocean Acidification on Fish," *American Journal of Physiology: Regulatory, Integrative and Comparative Physiology* 307, no. 9 (2014): R1061–R1084, https://doi.org/10.1152/ajpregu.00064.2014; Sean D. Connell et al., "The Other Ocean Acidification Problem: CO_2 as a Resource Among Competitors for Ecosystem Dominance," *Royal Society Publishing B: Biological Sciences* 368, no. 1627 (2013): 20120442, https://doi.org/10.1098/rstb.2012.0442; Rebecca Albright, "Reviewing the Effects of Ocean Acidification on Sexual Reproduction and Early Life History Stages of Reef-Building Corals," *Journal of Marine Sciences* 2011 (2011), https://doi.org/10.1155/2011/473615.

32. Gretchen E. Hofmann et al., "The Effect of Ocean Acidification on Calcifying Organisms in Marine Ecosystems: An Organism-to-Ecosystem Perspective," *Annual Review of Ecology, Evolution, and Systematics* 41 (2010): 127–147, https://doi.org/10.1146/annurev.ecolsys.110308.120227; James P. Barry, Stephen Widdicombe, and Jason M. Hall-Spencer, "Impacts of Ocean Acidification on Marine Biodiversity," in *Ocean Acidification*, ed. Jean-Pierre Gattuso and Lina Hansson (Oxford: Oxford University Press, 2020), 192–209.

33. Ximing Guo et al., "Diversity and Evolution of Living Oysters," *Journal of Shellfish Research* 37, no. 4 (2018): 755–771, https://doi.org/10.2983/035.037.0407.

34. Ute Eberle, "Running Boot Camps for Oysters to Train Them for a Warming World," Sea Grant California, August 17, 2022, https://caseagrant.ucsd.edu/news/running-boot-camps-oysters-train-them-warming-world.

35. Wendel W. Raymond et al., "Assessment of the Impacts of an Unprecedented Heatwave on Intertidal Shellfish of the Salish Sea," *Ecology* 103, no. 10 (2022): 1–7, https://doi.org/10.1002/ecy.3798.

5. THE ABUNDANT OCEAN

1. "Heʻeia Fishpond," Nā Kilo Honua o Heʻeia: Heʻeia Coastal Ocean Observing System, accessed April 28, 2023, http://www.nakilohonuaoheeia.org/site-description; Daniel McCoy et al., "Large-Scale Climatic Effects on Traditional Hawaiian Fishpond Aquaculture," *PLoS One* 12, no. 11 (2017): e0187951, https://doi.org/10.1371/journal.pone.0187951.
2. Erica Gies, "Hawaiʻi's Ancient Aquaculture Revival," *bioGraphic*, June 12, 2019, www.biographic.com/hawaiis-ancient-aquaculture-revival/.
3. McCoy et al., "Large-Scale Climatic Effects."
4. "Paepae o Heʻeia: Growing Seafood for Our Community One Pohaku at a Time," Paepae o Heʻeia, accessed April 28, 2023, www.paepaeoheeia.org/.
5. Kawika B. Winter et al., "Collaborative Research to Inform Adaptive Comanagement: A Framework for the Heʻeia National Estuarine Research Reserve," *Ecology and Society* 25, no. 4 (2020), https://doi.org/10.5751/ES-11895-250415.
6. Sam ʻOhu Gon and Kawika Winter, "A Hawaiian Renaissance That Could Save the World," *American Scientist* 107, no. 4 (2019): 232, https://doi.org/10.1511/2019.107.4.232.
7. Tsuneo Nakajima et al., "Common Carp Aquaculture in Neolithic China Dates Back 8,000 Years," *Nature Ecology & Evolution* 3, no. 10 (2019): 1415–1418, https://doi.org/10.1038/s41559-019-0974-3.
8. "Budj Bim Cultural Landscape," United Nations Educational, Scientific, and Cultural Organization World Heritage Convention, accessed April 28, 2023, https://whc.unesco.org/en/list/1577/.
9. Nakajima et al., "Common Carp Aquaculture."
10. Paul Greenberg, *Four Fish: The Future of the Last Wild Food* (New York: Penguin, 2010), 69–70.
11. Neil Ridler et al., "Integrated Multi-Trophic Aquaculture (IMTA): A Potential Strategic Choice for Farmers," *Aquaculture Economics & Management* 11, no. 1 (2007): 99–110, https://doi.org/10.1080/13657300701202767.
12. David Swanson, "A New Estimate of the Hawaiian Population for 1778, the Year of First European Contact," *Hūlili: Multidisciplinary Research on Hawaiian Well-Being* 11, no. 2 (2019): 203, https://doi.org/10.37712/hulili.2019.11-2.06.
13. Natalie Kurashima, Lucas Fortini, and Tamara Ticktin, "The Potential of Indigenous Agricultural Food Production Under Climate Change in Hawaiʻi," *Nature Sustainability* 2, no. 3 (2019): 191–199, https://doi.org/10.1038/s41893-019-0226-1.
14. Gies, "Hawaiʻi's Ancient Aquaculture Revival."
15. McCoy et al., "Large-Scale Climatic Effects."
16. Thomas T. Huxley, *Inaugural Meeting of the Fishery Congress* (London: William Clowes, 1883), 2–19.
17. "Marine Protected Areas: Preserving America's Oceans and Coasts," National Oceanic and Atmospheric Administration National Marine Protected Areas Center, accessed April 28, 2023, https://marineprotectedareas.noaa.gov/.
18. Peter J. S. Jones and Elizabeth M. De Santo, "The Race for Vast Remote 'Marine Protected Areas' May Be a Diversion," *Conversation*, September 1, 2016, https://theconversation.com/the-race-for-vast-remote-marine-protected-areas-may-be-a-diversion-64705.

19. Angelo O'Connor Villagomez, "Practicing My Culture Redux," The Saipan Blog, March 20, 2008, http://www.taotaotasi.com/2008/03/practicing-my-culture-redux.html.
20. Richard P. Duncan, Alison G. Boyer, and Tim M. Blackburn, "Magnitude and Variation of Prehistoric Bird Extinctions in the Pacific," *Proceedings of the National Academy of Sciences* 110, no. 16 (2013): 6436–6441, https://doi.org/10.1073/pnas.1216511110.
21. Full quote from Madi Williams: "The environment was extremely rich in resources and would have seemed limitless. Aotearoa New Zealand would have been covered in forest at this point and it possessed a range of birds and sea mammals unused to human predation. Over time, there was a reasonably fast reduction in these resources, and therefore horticulture began to be adopted on a large scale. This was a pattern that occurred in nearly all of Polynesia, but is exemplified most visibly in Aotearoa New Zealand. It was soon understood that the resources were finite and efforts were made to conserve them. Despite these efforts, the early inhabitants did make a significant impact on their new environments. The introduction of the kurī and kiore to Aotearoa New Zealand, along with hunting by humans, contributed to the extinction of some bird species and other vertebrates. One of the most extreme effects of the Polynesians on the Aotearoa New Zealand environment was deforestation. When European settlement occurred in the nineteenth century, around half the forest had already been lost. This begs the question, was deforestation a purposeful process or accidental? It has been suggested that it was purposeful and was intended to encourage the growth of non-cultivated edible plants after the effects of over-exploitation began to be felt. Another possible reason for deforestation was hunting of birds. The most probable cause of deforestation in Aotearoa New Zealand is burning by Polynesians. While there has been some debate concerning the extent of human impact versus natural occurrences, such as changes in climate, it is now widely accepted that human-induced fire was used to clear the forests, with the caveat that natural factors may have played a role." Madi Williams, *Polynesia, 900–1600* (Yorkshire: Art Humanities Press, 2021), 51.
22. Robert E. Johannes, "Traditional Marine Conservation Methods in Oceania and Their Demise," *Annual Review of Ecology and Systematics* 9, no. 1 (1978): 349–364, https://doi.org/10.1146/annurev.es.09.110178.002025.
23. Johannes, "Traditional Marine Conservation Methods."
24. Carlotta Leon Guerrero, "The U.N. Says 1 Million Species Could Disappear. Pacific Islands Have a Solution," Grist, May 8, 2018, www.grist.org/article/the-u-n-says-1-million-species-could-disappear-pacific-islands-have-a-solution/.
25. David A. Kroodsma et al., "Tracking the Global Footprint of Fisheries," *Science* 359, no. 6378 (February 2018): 904–908, https://doi.org/10.1126/science.aao5646.
26. 'Ohu Gon and Winter, "A Hawaiian Renaissance."
27. Sarah Wurz, "The Transition to Modern Behavior," *Nature Education Knowledge* 3, no. 10 (2012): 15, www.nature.com/scitable/knowledge/library/the-transition-to-modern-behavior-86614339/.
28. Francesco d'Errico et al., "Additional Evidence on the Use of Personal Ornaments in the Middle Paleolithic of North Africa," *Proceedings of the National Academy of Sciences* 6, no. 38 (2009): 16051–16056, https://doi.org/10.1073/pnas.0903532106.

29. Sally Mcbrearty and Alison S. Brooks, "The Revolution That Wasn't: A New Interpretation of the Origin of Modern Human Behavior," *Journal of Human Evolution* 39, no. 5 (November 2000): 453–463, https://doi.org/10.1006/jhev.2000.0435.
30. Sue O'Connor, interview with the authors, January 2022.
31. Sue O'Connor, Rintaro Ono, and Chris Clarkson, "Pelagic Fishing at 42,000 Years Before the Present and the Maritime Skills of Modern Humans," *Science* 334, no. 6059 (2011): 1117–1121, https://doi.org/10.1126/science.1207703.
32. In email correspondence with us (2022), Sue O'Connor told us that the original name they had used for the cave was incorrect, and the correct name is Asitau Kuru.
33. O'Connor, interview with the authors.
34. Masaki Fujita et al., "Advanced Maritime Adaptation in the Western Pacific Coastal Region Extends Back to 35,000–30,000 Years Before Present," *Proceedings of the National Academy of Sciences* 113, no. 40 (2016): 11184–11189, https://doi.org/10.1073/pnas.1607857113.
35. Sue O'Connor et al., "Fishing in Life and Death: Pleistocene Fish-Hooks from a Burial Context on Alor Island, Indonesia," *Antiquity* 91, no. 360 (2017): 1451–1468, https://doi.org/10.15184/aqy.2017.186.
36. Mark Kurlansky, *Cod: A Biography of the Fish That Changed the World* (New York: Penguin, 1997), 19–28.
37. "Climate Variability & Marine Fisheries," National Oceanic and Atmospheric Administration Pacific Fisheries Environmental Laboratory, accessed May 1, 2023, https://upwell.pfeg.noaa.gov/research/climatemarine/cmffish/cmffishery.html.
38. "The Surprising Story of Swordfish You May Not Know," *National Oceanic and Atmospheric Administration Fisheries*, April 30, 2019, www.fisheries.noaa.gov/feature-story/surprising-story-swordfish-you-may-not-know.
39. Moriaki Yasuhara and Curtis A. Deutsch, "Paleobiology Provides Glimpses of Future Ocean," *Science* 375, no. 6576 (2022): 25–26, https://doi.org/10.1126/science.abn2384.
40. Renato Salvatteci et al., "Smaller Fish Species in a Warm and Oxygen-Poor Humboldt Current System," *Science* 375, no. 6576 (January 2022): 101–104, https://doi.org/10.1126/science.abj0270.
41. Debora Mackenzie, "Cod Crisis Forces Canada to Curb Fishing," *New Scientist*, March 7, 1992, https://www.newscientist.com/article/mg13318110-400-cod-crisis-forces-canada-to-curb-fishing.
42. Jason Link et al., "The Effects of Area Closures on Georges Bank," *American Fisheries Society Symposium* 41 (January 2005): 345–368, https://www.researchgate.net/publication/263887097_The_Effects_of_Area_Closures_on_Georges_Bank.
43. Kurlansky, *Cod*, 186.
44. Kenneth F. Drinkwater, "The Response of Atlantic Cod (*Gadus morhua*) to Future Climate Change," *International Council for the Exploration of the Sea Journal of Marine Science* 62, no. 7 (2005): 1327–1337, https://doi.org/10.1016/j.icesjms.2005.05.015.
45. Michael L. Weber, *From Abundance to Scarcity: A History of U.S. Marine Fisheries Policy* (Washington, DC: Island Press, 2002), xv–xvii.
46. Kroodsma et al., "Tracking the Global Footprint of Fisheries."
47. "International Convention for the Safety of Life at Sea (SOLAS)," conclusion date: January 11, 1974, *United Nations Treaty Series Online*, registration no. I-18961, www.imo.org/en/About/Conventions/Pages/International-Convention-for-the-Safety-of-Life-at-Sea-(SOLAS),-1974.aspx.

48. Ricardo O. Amoroso et al., "Comment on 'Tracking the Global Footprint of Fisheries,'" *Science* 361, no. 6404 (2018): eaat6713, https://doi.org/10.1126/science.aat6713.
49. Boris Worm et al., "Rebuilding Global Fisheries," *Science* 325, no. 5940 (2009): 578–585, https://doi.org/10.1126/science.1173146.
50. Malin L. Pinsky et al., "Unexpected Patterns of Fisheries Collapse in the World's Oceans," *Proceedings of the National Academy of Sciences* 108, no. 20 (2011): 8317–8322, https://doi.org/10.1073/pnas.1015313108.
51. Malin L. Pinsky et al., "Marine Taxa Track Local Climate Velocities," *Science* 341, no. 6151 (September 2013): 1239–1242, https://doi.org/10.1126/science.1239352.
52. Summer Flounder Annual Catch Limit (ACL) is regulated by the National Marine Fisheries Service (NMFS), National Oceanographic and Atmospheric Administration (NOAA). States can transfer portions of their set quotas to other states as part of the management plan. "Final Rule for 2022 and Projected 2023 Summer Flounder, Scup, and Black Sea Bass Specifications," National Oceanic and Atmospheric Administration Fisheries, last modified December 22, 2021, www.fisheries.noaa.gov/action/final-rule-2022-and-projected-2023-summer-flounder-scup-and-black-sea-bass-specifications.
53. Data from Malin Pinsky's research showing these trends are illustrated beautifully in a Reuters visualization. Catch quotas presented here are from NOAA Fisheries. Maurice Tamman et al., "Undersea Science: The Planet's Hidden Climate Change," Reuters, accessed April 28, 2023, https://www.reuters.com/investigates/special-report/ocean-shock-warming/.
54. Malin L. Pinsky et al., "Fish and Fisheries in Hot Water: What Is Happening and How Do We Adapt?," *Population Ecology* 63, no. 1 (January 2021): 17–26, https://doi.org/10.1002/1438-390X.12050.
55. Pinsky et al., "Marine Taxa."
56. Tamman et al., "Undersea Science."
57. Julia Busiek, "Why's It Called El Niño, and How Did Scientists Figure Out What It Is?," Bay Nature, September 15, 2015, www.baynature.org/article/whys-it-called-el-nino-and-how-did-scientists-figure-out-what-it-is/.
58. "Pacific Decadal Oscillation," National Oceanic and Atmospheric Administration, National Centers for Environmental Information, last modified May 8, 2022, www.ncdc.noaa.gov/teleconnections/pdo/.
59. Rebecca Lindsey, "Climate Variability: Arctic Oscillation," National Oceanic and Atmospheric Administration, Climate.gov, last modified August 30, 2009, www.climate.gov/news-features/understanding-climate/climate-variability-arctic-oscillation.
60. Charles H. Greene and Andrew J. Pershing, "The Flip-Side of the North Atlantic Oscillation and Modal Shifts in Slope-Water Circulation Patterns," *Limnology and Oceanography* 48, no. 1 (January 2003): 319–322, https://doi.org/10.4319/lo.2003.48.1.0319.
61. Charles H. Greene et al., "Remote Climate Forcing of Decadal-Scale Regime Shifts in Northwest Atlantic Shelf Ecosystems," *Limnology and Oceanography* 58, no. 3 (May 2013): 803–816, https://doi.org/10.4319/lo.2013.58.3.0803803; Lindsey, "Climate Variability."
62. Penelope K. Hardy and Helen M. Rozwadowski, "Maury for Modern Times: Navigating a Racist Legacy in Ocean Science," *Oceanography* 33, no. 3 (2020): 10–15, https://doi.org/10.5670/oceanog.2020.302.
63. Hardy and Rozwadowski, "Maury for Modern Times."

64. Alexis P. Gumbs, *Undrowned: Black Feminist Lessons from Marine Mammals* (Chico, CA: AK Press, 2020), 103.
65. Jack Fritsch, "The Right Whale . . . but Wrong Story," Nantucket Historical Association, https://nha.org/research/nantucket-history/history-topics/the-right-whalebut-wrong-story/.
66. Ray Gambell, "International Management of Whales and Whaling: An Historical Review of the Regulation of Commercial and Aboriginal Subsistence Whaling," *Arctic* 46, no. 2 (1993): 97–107, https://doi.org/10.14430/arctic1330.
67. Sophie Monsarrat et al., "A Spatially Explicit Estimate of the Prewhaling Abundance of the Endangered North Atlantic Right Whale," *Conservation Biology* 30, no. 4 (August 2016): 1–9, https://doi.org/10.1111/cobi.12664.
68. Gambell, "International Management of Whales and Whaling."
69. "NARWC Annual Report Card," North Atlantic Right Whale Consortium Report Card, accessed April 28, 2023, www.narwc.org/report-cards.html.
70. Gumbs, *Undrowned*, 102.
71. Erin L. Meyer-Gutbrod et al., "Ocean Regime Shift Is Driving Collapse of the North Atlantic Right Whale Population," *Oceanography* 34, no. 3 (2021): 22–31, https://doi.org/10.5670/oceanog.2021.308.
72. Walter H. Munk, "On the Wind-Driven Ocean Circulation," *Journal of the Atmospheric Sciences* 7, no. 2 (1950): 80–93, https://doi.org/10.1175/1520-0469(1950)007<0080:OTWDOC>2.0.CO;2.
73. Stefan Rahmstorf et al., "Exceptional Twentieth-Century Slowdown in Atlantic Ocean Overturning Circulation," *Nature Climate Change* 5, no. 5 (2015): 475–480, https://doi.org/10.1038/nclimate2554.
74. Levke Caesar et al., "Observed Fingerprint of a Weakening Atlantic Ocean Overturning Circulation," *Nature* 556 (2018): 191–196, https://doi.org/10.1038/s41586-018-0006-5; David J. R. Thornalley et al., "Anomalously Weak Labrador Sea Convection and Atlantic Overturning During the Past 150 Years," *Nature* 556, no. 7700 (2018): 227–230, https://doi.org/10.1038/s41586-018-0007-4; Levke Caesar et al., "Current Atlantic Meridional Overturning Circulation Weakest in Last Millennium," *Nature Geoscience* 14, no. 3 (2021): 118–120, https://doi.org/10.1038/s41561-021-00699-z.
75. Charles Greene, Erin Meyer-Gutbrod, and Kimberley Davies, "How Climate Driven Ocean Changes Are Further Threatening North Atlantic Right Whales" (webinar, Lenfest Ocean Program, October 13, 2021), https://www.lenfestocean.org/en/news-and-publications/multimedia/webinar-on-how-climate-driven-ocean-changes-are-further-threatening-north-atlantic-right-whales.
76. Andrew J. Pershing et al., "Slow Adaptation in the Face of Rapid Warming Leads to Collapse of the Gulf of Maine Cod Fishery," *Science* 350, no. 6262 (2015): 809–812, https://doi.org/10.1126/science.aac9819.
77. Pershing et al., "Slow Adaptation in the Face of Rapid Warming."
78. Greene, Meyer-Gutbrod, and Davies, "How Climate Driven Ocean Changes."
79. Meyer-Gutbrod et al., "Ocean Regime Shift."
80. Brian D. Grieve et al., "Projecting the Effects of Climate Change on *Calanus finmarchicus* Distribution Within the U.S. Northeast Continental Shelf," *Scientific Reports* 7, no. 6264 (2017): 1–12, https://doi.org/10.1038/s41598-017-06524-1; Meyer-Gutbrod et al., "Ocean Regime Shift."

81. Grieve et al., "Projecting the Effects of Climate Change."
82. Beaugrand Gregory, Luczak Christophe, and Edwards Martin, "Rapid Biogeographical Plankton Shifts in the North Atlantic Ocean," *Global Change Biology* 15, no. 7 (July 2009): 1790–1803, https://doi.org/10.1111/j.1365-2486.2009.01848.x.
83. Michael J. Moore, *We Are All Whalers* (Chicago: University of Chicago Press, 2021), 114.
84. Meyer-Gutbrod et al., "Ocean Regime Shift"; Joshua D. Stewart et al., "Larger Females Have More Calves: Influence of Maternal Body Length on Fecundity in North Atlantic Right Whales," *Marine Ecology Progress Series* 689 (2022): 179–189, https://doi.org/10.3354/meps14040; Moore, *We Are All Whalers*, 154.
85. "NARWC Annual Report Card."
86. "North Atlantic Right Whale Calving Season 2021," National Oceanic and Atmospheric Administration National Centers Fisheries, February 12, 2021, https://www.fisheries.noaa.gov/feature-story/north-atlantic-right-whale-calving-season-2021.
87. "Unusual Mortality Events (UME's) are tracked by NOAA for multiple marine mammal species each year." Quote from "2017–2022 North Atlantic Right Whale Unusual Mortality Event," National Oceanic and Atmospheric Administration Fisheries, last modified April 4, 2023, www.fisheries.noaa.gov/national/marine-life-distress/2017-2022-north-atlantic-right-whale-unusual-mortality-event.
88. "Faces of Right Whale Conservation: Sofie Van Parijs, Acoustic Researcher," National Oceanic and Atmospheric Administration Fisheries, February 9, 2018, www.fisheries.noaa.gov/feature-story/faces-right-whale-conservation-sofie-van-parijs-acoustic-researcher
89. Erin Meyer-Gutbrod and Charles H. Greene, "Climate-Associated Regime Shifts Drive Decadal-Scale Variability in Recovery of North Atlantic Right Whale Population," *Oceanography* 27, no. 3 (2014): 148–153, https://doi.org/10.5670/oceanog.2014.64.
90. "BOEM Announces Multi-Agency Approach to Enhance Existing Protection Efforts for Endangered North Atlantic Right Whales," Bureau of Ocean Energy Management, February 7, 2022, https://www.boem.gov/newsroom/press-releases/boem-announces-multi-agency-approach-enhance-existing-protection-efforts.
91. Christopher M. Johnson et al., *Protecting Blue Corridors: Challenges and Solutions for Migratory Whales Navigating International and National Seas* (Washington, DC: Worldwide Wildlife Fund, 2022).
92. Eliza Oldach et al., "Managed and Unmanaged Whale Mortality in the California Current Ecosystem," *Marine Policy* 140 (June 2022): 105039, https://doi.org/10.1016/j.marpol.2022.105039.
93. William Sydeman, "Biological Response to Recent Changes in Climate" (presentation, Ocean Climate Summit at Fort Mason General's Residence, San Francisco, CA, May 17, 2016), http://climate.calcommons.org/aux/2016OceanClimateSummit/Web%20Update/Presentations/2_3_Sydeman.pdf.
94. Ryan M. McCabe et al., "An Unprecedented Coastwide Toxic Algal Bloom Linked to Anomalous Ocean Conditions," *Geophysical Research Letters* 43, no. 19 (October 2016): 10366–10376, https://doi.org/10.1002/2016GL070023.
95. "Commercial Dungeness Crab Season Opener Delayed and Commercial Rock Crab Season Closed," California Ocean Protection Council, accessed April 28, 2023, www.opc.ca.gov/2015/11/commercial-dungeness-crab-season-opener-delayed-and-commercial-rock-crab-season-closed/.

96. Jarrod A. Santora et al., "Habitat Compression and Ecosystem Shifts as Potential Links Between Marine Heatwaves and Record Whale Entanglements," *Nature Communications* 11, no. 1 (2020): 356, https://doi.org/10.1038/s41467-019-14215-w.
97. Jameal F. Samhouri et al., "Marine Heatwave Challenges Solutions to Human–Wildlife Conflict," *Proceedings of the Royal Society B: Biological Sciences* 288, no. 1964 (2021): 20211607, https://doi.org/10.1098/rspb.2021.1607.
98. Pacific Coast Federation of Fishermen's Associations, Inc. v. Chevron Corp, SHER, CGC-18-571285 (Cal. Super. Ct. 2018), http://climatecasechart.com/wp-content/uploads/sites/16/case-documents/2018/20181114_docket-CGC-18-571285_complaint.pdf.
99. "Climate Change Litigation Databases," Sabin Center for Climate Change Law, accessed April 28, 2023, http://climatecasechart.com/.
100. Kirk Moore, "Slinky Pots and Ropeless Gear: Next Angles for Whale Avoidance," *National Fisherman*, November 20, 2021, www.nationalfisherman.com/west-coast-pacific/slinky-pots-and-ropeless-gear-next-angles-for-whale-avoidance.
101. Jessica Morten et al., "Evaluating Adherence with Voluntary Slow Speed Initiatives to Protect Endangered Whales," *Frontiers in Marine Science* 9 (2022): 833206, https://doi.org/10.3389/fmars.2022.833206.
102. "Reducing the Risk of Ship Strikes on Endangered Whales," Greater Farallones Association, accessed April 28, 2023, https://farallones.org/reduce-whale-strikes/.
103. Matt, "After Two Whale Entanglements, Most Dungeness Crab Fishing Off California Will End Early," *SFist*, March 26, 2022, https://sfist.com/2022/03/26/after-two-whale-entanglements-most-dungeness-crab-fishing-off-california-will-end-early/.
104. Eliza Oldach et al., "Managed and Unmanaged Whale Mortality in the California Current Ecosystem," *Marine Policy* 140 (June 2022): 105039, https://www.sciencedirect.com/science/article/pii/S0308597X22000860#bib27.
105. "Hypoxia off the Pacific Northwest Coast," Partnership for Interdisciplinary Studies of Coastal Oceans, accessed April 28, 2023, https://www.piscoweb.org/sites/default/files/pdf/hypoxia_general%20low-res.pdf.
106. "J-Scope Forecast System," J-Scope, accessed April 28, 2023, http://www.nanoos.org/products/j-scope/forecasts.php.
107. Steve Lundeberg, "Ocean Hypoxia off Pacific Northwest Coast More Troubling Than Ever, Scientists Say," *Oregon State University*, September 08, 2021, https://today.oregonstate.edu/news/ocean-hypoxia-pacific-northwest-coast-more-troubling-ever-experts-say.
108. Julia Rosen, "A New Tool May Help Crab Fishers Sidestep Dead Zones," *Smithsonian Magazine*, March 8, 2022, https://www.smithsonianmag.com/innovation/new-tool-may-help-crab-fishers-sidestep-dead-zones-180979688.
109. Jock C. Currie et al., "Long-Term Change of Demersal Fish Assemblages on the Inshore Agulhas Bank Between 1904 and 2015," *Frontiers in Marine Science* 7 (2020): 355, https://doi.org/10.3389/fmars.2020.00355.
110. Helen Swingler, "PhD Researcher Recreates 111-Year-Old Fishing Survey," *University of Cape Town News*, July 10, 2020, www.news.uct.ac.za/article/-2020-07-10-phd-researcher-recreates-111-year-old-fishing-survey.
111. Kristina M. Barclay and Lindsey R. Leighton, "Predation Scars Reveal Declines in Crab Populations Since the Pleistocene," *Frontiers in Ecology and Evolution* 10 (2022): 810069, https://doi.org/10.3389/fevo.2022.810069.

112. Jon M. Erlandson and Torben C. Rick, "Archaeology Meets Marine Ecology: The Antiquity of Maritime Cultures and Human Impacts on Marine Fisheries and Ecosystems," *Annual Review of Marine Science* 2 (2010): 231–251, https://doi.org/10.1146/annurev.marine.010908.163749.

6. THE OPEN OCEAN

1. "San Francisco Community Welcomes Hikianalia and Crew," Hōkūle'a, September 16, 2018, https://www.hokulea.com/san-francisco-community-welcomes-hikianalia-and-crew/.
2. "Hikianalia Arrival Ceremony & Celebration in San Francisco," Hawai'i Chamber of Commerce of Northern California, December 24, 2018, https://www.hccnc.org/blog/2018/12/24/hikianalia-arrival-ceremony-amp-celebration-in-san-francisco.
3. Nainoa Thompson and Lehua Kamalu (presentation, Global Climate Action Summit, San Francisco, CA, September 15, 2018); Laura Guertin, "Ocean Sciences 2020—Plenary by Nainoa Thompson," *Advancing Earth and Science: Blogosphere*, February 17, 2020, https://blogs.agu.org/geoedtrek/2020/02/17/ocean-sciences-nainoa-thompson/.
4. Michael J. Morwood et al., "Fission-Track Ages of Stone Tools and Fossils on the East Indonesian Island of Flores," *Nature* 392 (1998): 173–176, https://doi.org/10.1038/32401.
5. Peter C. Van Welzen, John A. N. Parnell, and Johan W. Ferry Slik, "Wallace's Line and Plant Distributions: Two or Three Phytogeographical Areas and Where to Group Java?," *Biological Journal of the Linnean Society* 103, no. 3 (2011): 531–545, https://doi.org/10.1111/j.1095-8312.2011.01647.x.
6. Shimona Kealy et al., "Forty-Thousand Years of Maritime Subsistence Near a Changing Shoreline on Alor Island (Indonesia)," *Quaternary Science Reviews* 249 (2020): 106599, https://doi.org/10.1016/j.quascirev.2020.106599.
7. Alice B. Kehoe, *Traveling Prehistoric Seas: Critical Thinking on Ancient Transoceanic Voyages* (London: Routledge, 2016), 35.
8. Elizabeth Matisoo-Smith, "Ancient DNA and the Human Settlement of the Pacific: A Review," *Journal of Human Evolution* 79 (2015): 93–104, https://doi.org/10.1016/j.jhevol.2014.10.017.
9. Jon M. Erlandson, "Coastlines, Marine Ecology, and Maritime Dispersals in Human History," *Human Dispersal and Species Movement: From Prehistory to the Present* (2017): 147–163, https://doi.org/10.1017/9781316686942.007. Including this quote, "The widespread belief that coastal adaptations were a post-Pleistocene phenomenon had significant effects on archaeological perceptions of the importance of aquatic resources in human evolution, the role of coastlines in human dispersals, and the antiquity of human impacts on marine ecosystems."
10. Kehoe, *Traveling Prehistoric Seas*, 33.
11. Yuval Noah Harari, *Sapiens: A Brief History of Humankind* (New York: Harper, 2015), 64.
12. Les Groube et al., "A 40,000-Year-Old Human Occupation Site at Huon Peninsula, Papua New Guinea," *Nature* 324 (1986): 353–355, https://doi.org/10.1038/324453a0; Rintaro Ono, Alfred Pawlik, and Riczar Fuentes, "Island Migration, Resource Use, and Lithic Technology by Anatomically Modern Humans in Wallacea," in *Pleistocene Archaeology—Migration, Technology, and Adaptation*, ed. Rintaro Ono and Alfred Pawlik (London: IntechOpen, 2020), 1–204; Jon M. Erlandson and Torben C. Rick, "Archaeology

Meets Marine Ecology: The Antiquity of Maritime Cultures and Human Impacts on Marine Fisheries and Ecosystems," *Annual Review of Marine Science* 2 (2010): 231–251, https://doi.org/10.1146/annurev.marine.010908.163749; Japanese Archipelago Human Population Genetics Consortium, "The History of Human Populations in the Japanese Archipelago Inferred from Genome-Wide SNP Data with a Special Reference to the Ainu and the Ryukyuan Populations," *Nature* 57 (2012): 787–795, https://www.nature.com/articles/jhg2012114.

13. Fen Montaigne, "The Fertile Shore," *Smithsonian Magazine*, January 2020, https://www.smithsonianmag.com/science-nature/how-humans-came-to-americas-180973739/.
14. Timothy M. Rieth and Terry L. Hunt, "A Radiocarbon Chronology for Sāmoan Prehistory," *Journal of Archaeological Science* 35, no. 7 (July 2008): 1901–1927, https://doi.org/10.1016/j.jas.2007.12.001.
15. Kehoe, *Traveling Prehistoric Seas*, 41, 63.
16. Álvaro Montenegro, Richard T. Callaghan, and Scott M. Fitzpatrick, "Using Seafaring Simulations and Shortest-Hop Trajectories to Model the Prehistoric Colonization of Remote Oceania," *Proceedings of the National Academy of Sciences* 133, no. 45 (2016): 12685–12690, https://doi.org/10.1073/pnas.1612426113; Nicholas St. Fleur, "How Ancient Humans Reached Remote South Pacific Islands," *New York Times*, November 1, 2016, https://www.nytimes.com/2016/11/02/science/south-pacific-islands-migration.html.
17. Kehoe, *Traveling Prehistoric Seas*, 41, 63, 67.
18. Janet M. Wilmshurst et al., "High-Precision Radiocarbon Dating Shows Recent and Rapid Initial Human Colonization of East Polynesia," *Proceedings of the National Academy of Sciences* 108, no. 5 (2010): 1815–1820, https://doi.org/10.1073/pnas.1015876108. Including this quote, "In a meta-analysis of 1,434 radiocarbon dates from the region, reliable short-lived samples reveal that the colonization of East Polynesia occurred in two distinct phases: earliest in the Society Islands A.D. ~1025–1120, four centuries later than previously assumed; then after 70–265 y, dispersal continued in one major pulse to all remaining islands A.D. ~1190–1290."
19. Kehoe, *Traveling Prehistoric Seas*, 41, 63.
20. California Department of Parks and Recreation, "Malibu Lagoon Chumash-Polynesian Connection," California State Parks, accessed April 30, 2023, https://www.parks.ca.gov/?page_id=24433.
21. Nicholas Thomas, *Islanders: The Pacific in the Age of Empire* (New Haven, CT: Yale University Press, 2012), 11–12.
22. Madi Williams, *Polynesia, 900–1600* (Yorkshire: Arc Humanities Press, 2021), 1–104; Kehoe, *Traveling Prehistoric Seas*, 1–217.
23. Kamehameha Schools, *Kealaikahiki—Sea Road Between Hawai'i and Kahiki* (Ka'iwakiloumoku: He Kama Na Kahiki Symposium, 2020).
24. Kamehameha Schools, *Kealaikahiki*.
25. Anne Salmond, "Their Body Is Different, Our Body Is Different: European and Tahitian Navigators in the 18th Century," *History and Anthropology* 16, no. 2 (2005): 167–186, https://doi.org/10.1080/02757200500116105.
26. Salmond, "Their Body Is Different, Our Body Is Different."
27. Lars Eckstein and Anja Schwarz, "The Making of Tupaia's Map: A Story of the Extent and Mastery of Polynesian Navigation, Competing Systems of Wayfinding on James

Cook's Endeavour, and the Invention of an Ingenious Cartographic System," *Journal of Pacific History* 54, no. 1 (2019): 29, https://doi.org/10.1080/00223344.2018.1512369.

28. Salmond, "Their Body Is Different, Our Body Is Different."
29. Eckstein and Schwarz, "The Making of Tupaia's Map." For example, a Raiatean chant recorded by a missionary in 1817 shows directions from Tahiti to Hawai'i in terms of island outlines, astronomical observations, birds, and fish. See J. Orsmond and T. Henry, "The Birth of New Lands, After the Creation of Havai'i (Raiatea)," *Journal of the Polynesian Society* 3 (1894): 136–138.
30. Salmond, "Their Body Is Different, Our Body Is Different."
31. Epeli Hau'ofa, "Our Sea of Islands," *Contemporary Pacific* 6, no. 1 (1994): 148–161, https://doi.org/10.1515/9780824865542-005.
32. Ben Finney, "Myth, Experiment, and the Reinvention of Polynesian Voyaging," *American Anthropologist* 93, no. 2 (1991): 384–404, https://doi.org/10.1525/aa.1991.93.2.02a00060.
33. Hau'ofa, "Our Sea of Islands."
34. Herb Kawainui Kāne, "A Canoe Helps Hawaii Recapture Her Past," *National Geographic Magazine*, April 1976, 475.
35. Kehoe, *Traveling Prehistoric Seas*, 42.
36. Ocean Elders, "Pwo Navigator Nainoa Thompson Joins Dive in with Liz and Sylvia," YouTube, February 10, 2022, https://www.youtube.com/watch?v=iH5vguVTDR0.
37. Ocean Elders, "Pwo Navigator Nainoa Thompson."
38. Thompson and Kamalu, presentation.
39. "Review of Maritime Transport 2022," United Nations Conference on Trade and Development, accessed April 30, https://unctad.org/rmt2022.
40. Thompson and Kamalu, presentation.
41. Jennifer Ruesink (presentation, University of California, Davis, Bodega Marine Laboratory, Bodega, CA, December 14, 2019); Jennifer Ruesink, email message to author, June 22, 2022.
42. International Union for Conservation of Nature, "Marine Plastic Pollution," Issues Brief, November 2021, https://www.iucn.org/resources/issues-briefs/marine-plastic-pollution.
43. Lauren Roman et al., "Plastic Ingestion Is an Underestimated Cause of Death for Southern Hemisphere Albatrosses," *Conservation Letters* 14, no. 3 (2020): e12785, https://doi.org/10.1111/conl.12785.
44. Alan J. Jamieson et al., "Microplastics and Synthetic Particles Ingested by Deep-Sea Amphipods in Six of the Deepest Marine Ecosystems on Earth," *Royal Society Open Science* 6, no. 2 (2019): 180667, https://doi.org/10.1098/rsos.180667; Bonnie M. Hamilton et al., "Microplastics Around an Arctic Seabird Colony: Particle Community Composition Varies Across Environmental Matrices," *Science of the Total Environment* 773, no. 15 (2021): 145536, https://doi.org/10.1016/j.scitotenv.2021.145536.
45. Hannah Ritchie, "Where Does the Plastic in Our Oceans Come From?," Our World in Data, accessed April 25, 2023, https://ourworldindata.org/ocean-plastics.
46. "Journey to the Pacific Ocean Garbage Patch," Scripps Institution of Oceanography, August 7, 2009, https://scripps.ucsd.edu/news/journey-pacific-ocean-garbage-patch.
47. Annie Reisewitz and Christopher L. Clark, "Brave New Media World: A Science Communications Voyage to the Great Pacific Garbage Patch," abstract, American Geophysical Union (December 2009): https://ui.adsabs.harvard.edu/abs/2009AGUFMED14A..04R/abstract.

48. Laurent Lebreton et al., "Evidence That the Great Pacific Garbage Patch Is Rapidly Accumulating Plastic," *Scientific Reports* 8, no. 1 (2018): 1–15, https://doi.org/10.1038/s41598-018-22939-w.
49. Molly Taft, "Why Trying to Clean Up All the Ocean Plastic Is Pointless," Gizmodo, November 26, 2021, https://www.gizmodo.com/why-trying-to-clean-up-all-the-ocean-plastic-is-pointle-1848111529.
50. Lebreton et al., "Evidence That the Great Pacific Garbage Patch."
51. Lucy C. Woodall et al., "The Deep Sea Is a Major Sink for Microplastic Debris," *Royal Society Open Science* 1, no. 4 (2014): 140317, https://doi.org/10.1098/rsos.140317.
52. Linsey E. Haram et al., "Emergence of a Neopelagic Community Through the Establishment of Coastal Species on the High Seas," *Nature Communications* 12, no. 1 (2021): 8665, https://doi.org/10.1038/s41467-021-27188-6.
53. Chelsea Rochman et al., "Think Global, Act Local: Local Knowledge Is Critical to Inform Positive Change When It Comes to Microplastics," *Environmental Science and Technology* 55, no. 1 (2021): 4–6, https://doi.org/10.1021/acs.est.0c05746.
54. Roland Geyer, Jenna R. Jambeck, and Kara L. Lawet, "Production, Use, and Fate of All Plastics Ever Made," *Science Advances* 3, no. 7 (2017): e1700782, https://doi.org/10.1126/sciadv.1700782.
55. Taft, "Why Trying to Clean Up."
56. Waste Management and Pollution Control, "Pacific Leaders Call for Urgent Action on Marine Litter and Plastic Pollution," Secretariat of the Pacific Regional Environment Programme, September 14, 2021, https://www.sprep.org/news/pacific-leaders-call-for-urgent-action-on-marine-litter-and-plastic-pollution.
57. "PVS Crew Participate in Ocean Protection Summit and Sign Joint Oceans Declaration with French Polynesia," Hōkūleʻa, May 23, 2022, https://www.hokulea.com/pvs-crew-participate-in-ocean-protection-summit-and-sign-joint-oceans-declaration-with-french-polynesia/.
58. "Hōkūleʻa and Hikianalia Arrive in Tahiti," Hōkūleʻa, May 7, 2022, https://www.hokulea.com/hokulea-and-hikianalia-arrive-in-tahiti/.

7. THE POLAR WORLDS

1. Zachary M. Labe, "Climate Visualizations—Arctic: Sea-Ice Concentration/Extent," accessed April 24, 2023, https://zacklabe.com/arctic-sea-ice-extentconcentration/.
2. Jason Samenow and Kasha Patel, "It's 70 Degrees Warmer Than Normal in Eastern Antarctica. Scientists Are Flabbergasted," *Washington Post*, March 18, 2022, https://www.washingtonpost.com/weather/2022/03/18/antarctica-heat-wave-climate-change/.
3. Twila A. Moon et al., "The Expanding Footprint of Rapid Arctic Change," *Earth's Future* 7, no. 3 (March 2019): 212–218, https://doi.org/10.1029/2018EF001088.
4. Ivana Cvijanovic et al., "Future Loss of Arctic Sea-Ice Cover Could Drive a Substantial Decrease in California's Rainfall," *Nature Communications* 8, no. 1 (2017): 1947, https://doi.org/10.1038/s41467-017-01907-4; Jennifer Francis and Natasa Skific, "Evidence Linking Rapid Arctic Warming to Mid-Latitude Weather Patterns," *Philosophical Transactions of the Royal Society* 373, no. 2045 (2015): 20140170, https://doi.org/10.1098/rsta.2014.0170.

5. Bingyi Wu and Zhenkun Li, "Possible Impacts of Anomalous Arctic Sea Ice Melting on Summer Atmosphere," *International Journal of Climatology* 42, no. 3 (2022): 1818–1827, https://doi.org/10.1002/joc.7337.
6. Caitlyn Kennedy, "Earliest Satellite Images of Antarctica Reveal Highs and Lows for Sea Ice in the 1960s," National Oceanic and Atmospheric Administration Climate, last modified November 4, 2014, https://www.climate.gov/news-features/featured-images/earliest-satellite-images-antarctica-reveal-highs-and-lows-sea-ice.
7. Kelly Brunt and Alex Gardner, "Two Scientists Have a Frank and Honest Discussion About Antarctica: Kelly Brunt and Alex Gardner" (webinar, National Aeronautics and Space Administration, Ice, Cloud and Land Elevation Satellite-2, March 18, 2022), https://icesat-2.gsfc.nasa.gov/files/two-scientists-have-frank-and-honest-discussion-about-antarctica; "First Map of Antarctica's Moving Ice," National Aeronautics and Space Administration, Earth Observatory, accessed April 28, 2023, https://www.earthobservatory.nasa.gov/images/51781/first-map-of-antarcticas-moving-ice.
8. Thomas Frederikse et al., "The Causes of Sea-Level Rise Since 1900," *Nature* 584, no. 7821 (2020): 393–397, https://doi.org/10.1038/s41586-020-2591-3.
9. "Collapse of the Larsen-B Ice Shelf—World of Change," National Aeronautics and Space Administration, Earth Observatory, accessed April 28, 2023, https://www.earthobservatory.nasa.gov/world-of-change/LarsenB.
10. "U.S. Coastline to See Up to a Foot of Sea Level Rise by 2050," National Oceanic and Atmospheric Administration, February 15, 2022, https://www.noaa.gov/news-release/us-coastline-to-see-up-to-foot-of-sea-level-rise-by-2050; William V. Sweet et al., *2022: Global and Regional Sea Level Rise Scenarios for the United States: Updated Mean Projections and Extreme Water Level Probabilities Along U.S. Coastlines* (Silver Spring, MD: National Oceanic and Atmospheric Administration, 2022).
11. Elizabeth Rush, *Rising: Dispatches from the New American Shore* (Minneapolis, MN: Milkweed Editions, 2018), 62–67, 93–97, 162–166.
12. Rush, *Rising*, 14.
13. Elizabeth Rush, "First Passage: A Journey Toward Motherhood in the Age of Glacial Loss," *Orion Magazine*, June 2, 2021, https://www.orionmagazine.org/article/first-passage/.
14. "The Threat from Thwaites: The Retreat of Antarctica's Riskiest Glacier," Cooperative Institute for Research in Environmental Sciences, last modified December 13, 2021, https://www.cires.colorado.edu/news/threat-thwaites-retreat-antarctica's-riskiest-glacier.
15. Andrea Dutton and Kurt Lambeck, "Ice Volume and Sea Level During the Last Interglacial," *Science* 337, no. 6091 (2012): 216–219, https://doi.org/10.1126/science.1205749.
16. Andrea Dutton et al., "Sea-Level Rise Due to Polar Ice-Sheet Mass Loss During Past Warm Periods," *Science* 349, no. 6244 (2015): aaa4019, https://doi.org/10.1126/science.aaa4019.
17. Frederikse et al., "The Causes of Sea-level Rise Since 1900."
18. Andrea Dutton, "Reframing Sea Level Rise" (presentation, TEDxUF, Gainesville, FL, April 1, 2017), https://www.youtube.com/watch?v=OfpMO4WLDLE.
19. Hope Jahren, *The Story of More: How We Got to Climate Change and Where to Go from Here* (New York: Vintage, 2020), 151.
20. Richa Syal, "License to Krill: The Destructive Demand for a 'Better' Fish Oil," *Guardian*, September 7, 2021, https://www.theguardian.com/environment/2021/sep/07/license-to-krill-the-destructive-demand-for-a-better-fish-oil.

21. Gloria Dickie, "In Antarctica, Does a Burgeoning Krill Fishery Threaten Wildlife?," Reuters, February 24, 2022, https://www.reuters.com/business/cop/antarctica-does-burgeoning-krill-fishery-threaten-wildlife-2022-02-24/.
22. George M. Watters, Jefferson T. Hinke, and Christian S. Reiss, "Long-Term Observations from Antarctica Demonstrate That Mismatched Scales of Fisheries Management and Predator-Prey Interaction Lead to Erroneous Conclusions About Precaution," *Scientific Reports* 10, no. 1 (2020): 2314, https://doi.org/10.1038/s41598-020-59223-9; Lucas Krüger et al., "Antarctic Krill Fishery Effects Over Penguin Populations Under Adverse Climate Conditions: Implications for the Management of Fishing Practices," *Ambio* 50 (2021): 560–571, https://doi.org/10.1007/s13280-020-01386-w.
23. Luis A. Hückstädt, "Weddell Seal: *Leptonychotes weddellii*," in *Encyclopedia of Marine Mammals*, 3rd ed., ed. Bernd Würsig, Johannes G. M. Thewissen, and Kit M. Kovac (Cambridge, MA: Academic Press, 2018), 1048–1051.
24. Michael F. Cameron and Donald B. Siniff, "Age-Specific Survival, Abundance, and Immigration Rates of a Weddell Seal (*Leptonychotes weddellii*) Population in McMurdo Sound, Antarctica," *Canadian Journal of Zoology* 82, no. 4 (April 2004): 601–615, https://doi.org/10.1139/z04-025.
25. Michelle LaRue, "Many Hands Make Light Work: Crowdsourcing Reveals Population Status for Weddell Seals in Antarctica" (webinar, University of British Columbia, Vancouver, Canada, March 4, 2022), https://www.youtube.com/watch?v=W-NPRDWo-PI.
26. Michelle A. LaRue et al., "Satellite Imagery Can Be Used to Detect Variation in Abundance of Weddell Seals (*Leptonychotes weddellii*) in Erebus Bay, Antarctica," *Polar Biology* 34 (2011): 1727–1737, https://doi.org/10.1007/s00300-011-1023-0.
27. "Penguin Watch: About," Zooniverse, https://www.zooniverse.org/projects/penguintom79/penguin-watch/about/research; Henry F. Houskeeper et al., "Automated Satellite Remote Sensing of Giant Kelp at the Falkland Islands (Islas Malvinas)," *PLoS One* 17, no. 1 (2022): e0257933, https://doi.org/10.1371/journal.pone.0257933.
28. Michelle A. LaRue et al., "Engaging 'the Crowd' in Remote Sensing to Learn About Habitat Affinity of the Weddell Seal in Antarctica," *Remote Sensing in Ecology and Conservation* 6, no. 1 (March 2020): 70–78, https://doi.org/10.1002/rse2.124.
29. Michelle A. LaRue et al., "Insights from the First Global Population Estimate of Weddell Seals in Antarctica," *Science Advances* 7, no. 39 (2021): eabh3674, https://doi.org/10.1126/sciadv.abh3674.
30. Stacy L. Deppeler and Andrew T. Davidson, "Southern Ocean Phytoplankton in a Changing Climate," *Frontiers in Marine Science* 4 (2017): 40, https://doi.org/10.3389/fmars.2017.00040.
31. LaRue et al., "Insights from First Global Population Estimate."
32. Jacqueline M. Grebmeier, "Shifting Patterns of Life in the Pacific Arctic and Sub-Arctic Seas," *Annual Review of Marine Science* 4, no. 1 (2012): 63–78, https://doi.org/10.1146/annurev-marine-120710-100926; Donna D. W. Hauser, "Seasonal Sea Ice and Arctic Migrations of the Beluga Whale," *Alaska Park Science*, June 12, 2018, https://www.nps.gov/articles/aps-17-1-9.htm.
33. Donna D. W. Hauser, "Diverse Responses and Emerging Risks for Marine Mammals in a Rapidly Changing Arctic" (presentation, Arctic Research Consortium of the United

States, Fairbanks, Alaska, March 25, 2019), https://www.arcus.org/research-seminar-series/archive; Donna D. W. Hauser et al., "Indirect Effects of Sea Ice Loss on Summer-Fall Habitat and Behaviour for Sympatric Populations of an Arctic Marine Predator," *Diversity and Distributions* 24, no. 6 (2018): 791–799, https://doi.org/10.1111/ddi.12722.

34. Christophe Kinnard et al., "Reconstructed Changes in Arctic Sea Ice Over the Past 1,450 Years," *Nature* 479, no. 7374 (2011): 509–512, https://doi.org/10.1038/nature10581.
35. Kinnard et al., "Reconstructed Changes in Arctic Sea Ice."
36. David Docquier, "Will the Arctic Be Ice Free Earlier Than Previously Thought?," *European Geosciences Union Cryospheric Sciences Blog*, September 17, 2021, https://blogs.egu.eu/divisions/cr/2021/09/17/will-the-arctic-be-ice-free-earlier-than-previously-thought/.
37. Matthew O'Regan et al., "A Synthesis of the Long-Term Paleoclimatic Evolution of the Arctic," *Oceanography* 24, no. 3 (2011): 66–80, https://www.jstor.org/stable/24861300.
38. Russell Goldman, "Russian Tanker Completes Arctic Passage Without Aid of Icebreakers," *New York Times*, August 25, 2017, https://www.nytimes.com/2017/08/25/world/europe/russia-tanker-christophe-de-margerie.html.
39. Hauser, "Diverse Responses and Emerging Risks."
40. Enrico P. Metzner, Marc Salzmann, and Rüdiger Gerdes, "Arctic Ocean Surface Energy Flux and the Cold Halocline in Future Climate Projections," *Journal of Geophysical Research: Oceans* 125, no. 2 (2020): e2019JC015554, https://doi.org/10.1029/2019JC015554.
41. Igor V. Polyakov et al., "Greater Role for Atlantic Inflows on Sea-Ice Loss in the Eurasian Basin of the Arctic Ocean," *Science* 356, no. 6335 (2017): 285–291, https://doi.org/10.1126/science.aai82.
42. Rebecca A. Woodgate, "Increases in the Pacific Inflow to the Arctic from 1990 to 2015, and Insights into Seasonal Trends and Driving Mechanisms from Year-Round Bering Strait Mooring Data," *Progress in Oceanography* 190 (January 2018): 124–154, https://doi.org/10.1016/j.pocean.2017.12.007.
43. Maria Fossheim et al., "Recent Warming Leads to a Rapid Borealization of Fish Communities in the Arctic," *Nature Climate Change* 5, no. 7 (2015): 673–677, https://doi.org/10.1038/nclimate2647.
44. Henry P. Huntington et al., "Evidence Suggests Potential Transformation of the Pacific Arctic Ecosystem Is Underway," *Nature Climate Change* 10, no. 4 (2020): 342–348, https://doi.org/10.1038/s41558-020-0695-2.
45. National Parks Traveler Staff, "Seabird Die-Off Reported at Bering Land Bridge National Preserve," *National Parks Traveler*, September 10, 2019, https://www.nationalparkstraveler.org/2019/09/seabird-die-reported-bering-land-bridge-national-preserve.
46. "2018–2022 Ice Seal Unusual Mortality Event in Alaska," National Oceanic and Atmospheric Administration, Fisheries, last modified August 16, 2022, https://www.fisheries.noaa.gov/alaska/marine-life-distress/2018-2022-ice-seal-unusual-mortality-event-alaska.
47. Paula Dobbyn, "Vigilance Advised After Aaxitoxin Found in Clams," Alaska Sea Grant, November 19, 2019, https://www.alaskaseagrant.org/2019/11/19/vigilance-advised-after-saxitoxin-found-in-clams/; Yereth Rosen, "For the First Time, Scientists Have Found a Dangerous Toxin from Algae in the Bering Strait and Chukchi Sea," *Arctic Today*, December 12, 2019, https://www.arctictoday.com/for-the-first-time-scientists-have-found-a-dangerous-toxin-from-algae-in-the-bering-strait-and-chukchi-sea/.

48. Alex DeMarban, "Big Changes from Warming Climate Boost Risk of Toxic Algae Blooms off Alaska Coasts, Report Says," *Anchorage Daily News*, December 11, 2018, https://www.adn.com/alaska-news/2018/12/12/big-changes-from-warming-climate-boost-risk-of-toxic-algae-blooms-off-alaska-coasts-report-says/.
49. Hal Bernton, "Why Are Birds and Seals Starving in a Bering Sea Full of Fish?," *Seattle Times*, November 4, 2019, https://www.seattletimes.com/seattle-news/environment/why-are-birds-and-seals-starving-in-a-bering-sea-full-of-fish/.
50. SueEllen Campbell, "Key Articles Addressing Range of Changes in 'New Arctic,'" *Yale Climate Connections*, July 22, 2021, https://www.yaleclimateconnections.org/2021/07/key-articles-addressing-range-of-changes-in-new-arctic/; National Science Foundation's Navigating the New Arctic: requesting proposals that seek "innovations in fundamental convergence research across the social, natural, environmental, computing and information sciences, and engineering that address the interactions or connections among natural and built environments and social systems, and how these connections inform our understanding of Arctic change and its local and global effects." Quote from "Navigating the New Arctic (NNA)," National Science Foundation, last modified 2023, https://www.nsf.gov/geo/opp/arctic/nna/index.jsp.
51. "NSF's 10 Big Ideas," National Science Foundation, accessed April 27, 2021, https://www.nsf.gov/news/special_reports/big_ideas/index.jsp.
52. William E. Easterling et al., "Dear Colleague Letter: Stimulating Research Related to Navigating the New Arctic (NNA), One of NSF's 10 Big Ideas," National Science Foundation, last modified February 22, 2018, https://www.nsf.gov/pubs/2018/nsf18048/nsf18048.jsp.
53. Kaare Ray Sikuaq Erickson, "Successful Engagement Between Iñupiat and Scientists in Utqiaġvik, Alaska: A Sociocultural Perspective" (master's thesis, University of Alaska Anchorage, 2020).
54. Kaare Ray Sikuaq Erickson, "Indigenous Technology and Inuit History" (webinar, Arctic Research Consortium of the United States, May 27, 2021), https://www.arcus.org/research-seminar-series/archive.
55. Jenna Kunze, "As the Arctic Warms, the Inupiat Adapt," *High Country News*, July 1, 2020, https://www.hcn.org/issues/52.7/indigenous-affairs-climate-change-what-choice-do-we-have.
56. Emily Witt, "An Alaskan Town Is Losing Ground—and a Way of Life," *New Yorker*, November 21, 2022, https://www.newyorker.com/magazine/2022/11/28/an-alaskan-town-is-losing-ground-and-a-way-of-life.
57. David López-Carr and Jessica Marter-Kenyon, "Human Adaptation: Manage Climate-Induced Resettlement," *Nature* 517, no. 7534 (2015): 265–267, https://doi.org/10.1038/517265a.
58. Robin Bronen and F. Stuart Chapin III, "Adaptive Governance and Institutional Strategies for Climate-Induced Community Relocations in Alaska," *Proceedings of the National Academy of Sciences* 110, no. 3 (2013): 9320–9325, https://doi.org/10.1073/pnas.1210508111.
59. Stephen Sackur, "The Alaskan Village Set to Disappear Under Water in a Decade," British Broadcasting Corporation, July 30, 2013, https://www.bbc.com/news/magazine-23346370.
60. Moon et al., "The Expanding Footprint of Rapid Arctic Change."

61. Charles K. Paull, interview with authors at the Ocean Sciences Meeting, February 2020.
62. Charles K. Paull et al., "Rapid Seafloor Changes Associated with the Degradation of Arctic Submarine Permafrost," *Proceedings of the National Academy of Sciences* 119, no. 12 (2022): e2119105119, https://doi.org/10.1073/pnas.2119105119.
63. Kaare Ray Sikuaq Erickson and Tero Mustonen, "Increased Prevalence of Open Water During Winter in the Bering Sea: Cultural Consequences in Unalakleet, Alaska," *Oceanography* 35, no. 3–4 (2022): 180–188, https://doi.org/10.5670/oceanog.2022.135.
64. Melanie Lenart, "Haskell Receives $20 Million Climate Change Grant," *Native Science Report*, November 8, 2022, https://nativesciencereport.org/2022/11/haskell-indian-nations-university-receives-20-million-climate-change-grant/.
65. Aimee Huntington et al., "A First Assessment of Microplastics and Other Anthropogenic Particles in Hudson Bay and the Surrounding Eastern Canadian Arctic Waters of Nunavut," *FACETS* 5, no. 1 (2020): 432–454, https://doi.org/10.1139/facets-2019-0042; Melanie Bergmann et al., "Plastic Pollution in the Arctic," *Nature Reviews Earth and Environment* 3, no. 5 (2022): 323–337, https://www.nature.com/articles/s43017-022-00279-8.
66. Sheila Watt-Cloutier, *The Right to Be Cold* (Toronto: Penguin, 2018), 337.
67. Christopher Pearson et al., "Mercury and Trace Metal Wet Deposition Across Five Stations in Alaska: Controlling Factors, Spatial Patterns, and Source Regions," *Atmospheric Chemistry and Physics* 19, no. 10 (2019): 6913–6929, https://doi.org/10.5194/acp-19-6913-2019; Alexandra Steffen et al., "A Synthesis of Atmospheric Mercury Depletion Event Chemistry in the Atmosphere and Snow," *Atmospheric Chemistry and Physics* 8, no. 6 (2008): 1445–1482, https://doi.org/10.5194/acp-8-1445-2008.
68. France Collard and Amalie Ask, "Plastic Ingestion by Arctic Fauna: A Review," *Science of the Total Environment* 786 (2021): 147462, https://doi.org/10.1016/j.scitotenv.2021.147462.
69. S. Percy Smith, "Hawaiki: The Whence of the Maori: Being an Introduction to Rarotonga History: Part III," *Journal of the Polynesian Society* 8, no. 1 (1899): 10–11, https://www.jps.auckland.ac.nz/document//Volume_8_1899/Volume_8,_No._1,_March_1899/Hawaiki%3A_the_whence_of_the_Maori,_being_an_introduction_to_Rarotongan_history%3A_Part_III,_by_S._Percy_Smith,_p_1-48/p1.
70. Priscilla M. Wehi et al., "A Short Scan of Māori Journeys to Antarctica," *Journal of the Royal Society of New Zealand* 52, no. 5 (2022): 587–598, https://doi.org/10.1080/03036758.2021.1917633.
71. Michael Stevens, "Beyond the Southern Angle of the Polynesian Triangle: Māori Associations with the Southern Ocean and Antarctica," literature review of Te Tai Uka a Pia, *National Science Challenges: The Deep South*, February 2021, https://deepsouthchallenge.co.nz/resource/beyond-the-southern-angle-of-the-polynesian-triangle/.
72. Wehi et al., "A Short Scan of Māori Journeys."
73. Sandy Morrison and Aimee Kaio, "Te Tai Uka a Pia: Southern Ocean Voyaging," (webinar, Māori and Antarctica: Ka mua, ka muri Seminar Series 2020, July 27, 2020), https://maoriantarctica.org/voyaging-south/.
74. "Apollo Image Atlas: AS17-148-22727," Lunar and Planetary Institute, https://www.lpi.usra.edu/resources/apollo/frame/?AS17-148-22727.
75. "Finding Shackleton's Lost Ice Ship," Endurance22, accessed April 23, 2023, https://endurance22.org/.

76. Jon Copley, "Exploring the Antarctic Deep Seas Took Me Back in Time," *Popular Science*, November 7, 2017, https://www.popsci.com/explored-antarctic-deep-seas-for-blue-planet-ii/.
77. "Endurance Is Found," Endurance22, March 9, 2022, https://www.endurance22.org/endurance-is-found.

8. THE DEEP

1. Camilo Mora et al., "How Many Species Are There on Earth and in the Ocean?," *PLoS Biology* 9, no. 8 (2011): e1001127, https://doi.org/10.1371/journal.pbio.1001127.
2. "Lost Shipping Container Study," Monterey Bay Aquarium Research Institute, accessed April 30, 2023, https://www.mbari.org/project/lost-shipping-container-study/.
3. Oren T. Frey and Andrew P. DeVogelaere, *The Containerized Shipping Industry and the Phenomenon of Containers Lost at Sea* (Washington, DC: National Oceanic and Atmospheric Administration, 2014), 8.
4. Frey and DeVogelaere, *The Containerized Shipping Industry*.
5. "Restoration Plan for the M/V Med Taipei ISO Container Discharge Incident," Monterey Bay National Marine Sanctuary, accessed April 30, 2023, https://montereybay.noaa.gov/resourcepro/mt/welcome.html.
6. Alan J. Jamieson et al., "Microplastics and Synthetic Particles Ingested by Deep-Sea Amphipods in Six of the Deepest Marine Ecosystems on Earth," *Royal Society Open Science* 6, no. 2 (2019): 180667, https://doi.org/10.1098/rsos.180667.
7. Herman A. Karl et al., "Search for Containers of Radioactive Waste on the Sea Floor," in *Beyond the Golden Gate: Oceanography, Geology, Biology, and Environmental Issues in the Gulf of the Farallones*, ed. Herman A. Karl et al. (Reston, VA: U.S. Department of the Interior, 2001), 101–116; Veronika Kivenson et al., "Ocean Dumping of Containerized DDT Waste Was a Sloppy Process," *Environmental Science and Technology* 53, no. 6 (2019): 2971–2980, https://doi.org/10.1021/acs.est.8b05859.
8. Steven Bedard, "An Audacious Plan in the Twilight," *bioGraphic*, April 26, 2016, http://www.biographic.com/an-audacious-plan-in-the-twilight/.
9. Here's the Fish: Matt Pederson, "Bedazzling New Anthias—Pay No Attention to the Shark," *Reef to Rainforest*, September 25, 2018, http://www.reef2rainforest.com/2018/09/25/bedazzling-new-anthias-pay-no-attention-to-the-shark/.
10. California Academy of Sciences, "Sixgill Shark Sighting at 420 feet," YouTube, June 2, 2018, http://www.youtube.com/watch?v=pSZrmoEwRoQ.
11. Jon Sharman, "Divers 'Enchanted' by Colourful New Species of Reef Fish Fail to Spot Huge Shark Cruising Above Them," *Independent*, September 27, 2018, http://www.independent.co.uk/news/video-shark-over-group-species-reef-fish-brazil-atlantic-tosanoides-aphrodite-a8558311.html.
12. "Native Hawaiian Permission and Release Protocol for ROV Deep Dives in Papahānaumokuākea Marine National Monument," Office of Hawaiian Affairs, accessed April 28, 2023, http://www.19of32x2yl33s804xza0gf14-wpengine.netdna-ssl.com/wp-content/uploads/permission-and-release-protocol-ROV-deep-dive-expedition.pdf.
13. Huihui Kanahele-Mossman, "Building Relationships to Papahānaumokuākea Through Kanaka 'Ōiwi (Native Hawaiian) Oral Traditions" (webinar, National Oceanic and

Atmospheric Administration, February 17, 2022), https://sanctuaries.noaa.gov/education/teachers/building-relationships-to-papahanaumokuakea.html; Kanoe Morishige Native Hawaiian Specialist Contractor with Lynker in support of NOAA National Ocean Service, email to authors, April 3, 2023.

14. Jules Verne, "Pro and Con," in *Twenty Thousand Leagues Under the Sea* (France: Pierre-Jules Hetzel, 1869).
15. Thomas Anderson and Tony Rice, "Deserts on the Sea Floor: Edward Forbes and His Azoic Hypothesis for a Lifeless Deep Ocean," *Endeavour* 30, no. 4 (2006): 131–137, https://doi.org/10.1016/j.endeavour.2006.10.003.
16. John Isaac, "1857—Laying the Atlantic Telegraph Cable from Ship to Shore," Atlantic Cables, last modified February 16, 2022, https://atlantic-cable.com/Books/1857Isaac/.
17. Allison Marsh, "The First Transatlantic Telegraph Cable Was a Bold, Beautiful Failure," IEEE Spectrum, October 31, 2019, https://spectrum.ieee.org/the-first-transatlantic-telegraph-cable-was-a-bold-beautiful-failure.
18. Donard de Cogan, "Dr E. O. W. Whitehouse and the 1858 Trans-Atlantic Cable," Atlantic Cables, last modified January 10, 2017, https://atlantic-cable.com/Books/Whitehouse/DDC/index.htm.
19. Jeffreys J. Gwyn, "The Deep-Sea Dredging Expedition in H.M.S. 'Porcupine,'" *Nature* 1 (1869): 166–168, https://doi.org/10.1038/001166a0.
20. "Dive and Discover—The Challenger Expedition," Woods Hole Oceanographic Institution, accessed April 30, 2023, http://www.divediscover.whoi.edu/history-of-oceanography/the-challenger-expedition/.
21. C. Wyville Thompson and John Murray, *A Summary of the Scientific Results Obtained at the Sounding, Dredging and Trawling Stations of H.M.S. 1895 Challenger: Part 6*, (London: Eyre and Spottiswoode), 1: 1–796.
22. The "Zoology" section alone is thirty-two volumes in forty printed books. A single volume might cost you $50 to $100; collectors may find the full set available if they are willing to pay over $150,000.
23. "Marie Tharp," Woods Hole Oceanographic Institution, accessed April 28, 2023, http://www.whoi.edu/news-insights/content/marie-tharp/?pid=7500&tid=7342&cid=23306.
24. Sarah Fecht, "8 Surprising Facts About Marie Tharp, Mapmaker Extraordinaire," State of the Planet, July 29, 2020, http://news.climate.columbia.edu/2020/07/29/surprising-facts-marie-tharp/.
25. "Marie Tharp," Woods Hole Oceanographic Institution.
26. "Marie Tharp, Pioneering Oceanographer—Map of the Atlantic Ocean," University of Chicago Library, accessed April 29, 2023, http://www.lib.uchicago.edu/collex/exhibits/marie-tharp-pioneering-oceanographer/1957-atlantic-ocean-map/.
27. Suzanne O'Connell, "Marie Tharp Pioneered Mapping the Bottom of the Ocean 6 Decades Ago—Scientists Are Still Learning About Earth's Last Frontier," *Conversation*, February 28, 2022, https://theconversation.com/marie-tharp-pioneered-mapping-the-bottom-of-the-ocean-6-decades-ago-scientists-are-still-learning-about-earths-last-frontier-142451.
28. "Marie Tharp," Woods Hole Oceanographic Institution.
29. "Marie Tharp," Woods Hole Oceanographic Institution.
30. "World Ocean Floor," Library of Congress, accessed April 30, 2023, https://www.loc.gov/item/2010586277/.

250   8. THE DEEP

31. Andrew Revkin, "Two Planet-Mapping Scientists Explore the Storytelling Power of Cartography," YouTube, July 27, 2020, http://www.youtube.com/watch?v=wVNLVd6ZXH0.
32. Dawn Wright, "To Save Earth's Climate, Map the Oceans," *Esri*, Winter 2022, https://www.esri.com/about/newsroom/arcnews/to-save-earths-climate-map-the-oceans/.
33. One example is the Pacific Light Cable Network, owned by Google/Meta, which is 11,806 km (7,500 miles) from Philippines and Taiwan to Los Angeles: "Submarine Cable Map—Pacific Light Cable Network," TeleGeography, accessed April 25, 2023, http://www.submarinecablemap.com/submarine-cable/pacific-light-cable-network-plcn.
34. Wright, "To Save Earth's Climate."
35. Dawn J. Wright and Christian Harder, eds., *GIS for Science, Volume 2: Applying Mapping and Spatial Analytics* (Redlands, CA: Esri, 2020).
36. National Geographic Education Staff, "We're a Young Explorer and a Scientist, and These Are Our Ocean Stories. What's Yours?," *National Geographic*, September 2, 2021, http://www.blog.education.nationalgeographic.org/2021/09/02/were-a-young-explorer-and-a-scientist-and-these-are-our-ocean-stories-whats-yours/.
37. Owen Kibenge, "AAAS Fellow Dawn Wright Maps the Ocean Floor, Turbulent Times in Science," American Association for the Advancement of Science, August 26, 2020, http://www.aaas.org/membership/member-spotlight/aaas-fellow-dawn-wright-maps-ocean-floor-turbulent-times-science.
38. Revkin, "Two Planet-Mapping Scientists."
39. "1972—The Trail Gets Hot," Woods Hole Oceanographic Institution, accessed April 25, 2023, http://www.whoi.edu/feature/history-hydrothermal-vents/discovery/1972.html.
40. Robert Ballard, "Notes on a Major Oceanographic Find," *Oceanus* 20, no. 3 (Summer 1977): 35–44.
41. Susan E. Humphris, Christopher R. German, and J. Patrick Hickey, "Fifty Years of Deep Ocean Exploration with the DSV Alvin," *Eos*, June 3, 2014, https://eos.org/features/in-june-2014-the-deep-submergence-vehicle-dsv-alvin-the-worlds-first-deep-diving-sub-marine-dedicated-to-scientific-research-in-the-united-states-celebrated-its-50th-anniversary.
42. Ballard, "Notes on a Major Oceanographic Find."
43. "Diving to the Rosebud Vents—Galápagos Rift," Woods Hole Oceanographic Institution, accessed April 27, 2023, http://www.whoi.edu/press-room/news-release/diving-to-the-rosebud-vents-galapagos-rift/; "Dive and Discover—Hydrothermal Vents," Woods Hole Oceanographic Institution, accessed April 28, 2023, http://www.divediscover.whoi.edu/hydrothermal-vents/.
44. An historical time line can be found here, which includes scientific details but also mentions of submersible inspections and maintenance, impacts of federal budget cuts, and the overhaul required to extend diving to 6,500 meters: "History of *Alvin*," Woods Hole Oceanographic Institution, accessed 2022, http://www.whoi.edu/what-we-do/explore/underwater-vehicles/hov-alvin/history-of-alvin/.
45. "Alvin Upgrade," Woods Hole Oceanographic Institution, accessed 2023, https://www.whoi.edu/what-we-do/explore/underwater-vehicles/hov-alvin/history-of-alvin/alvin-upgrade/.
46. Charles Paull et al., "Biological Communities at the Florida Escarpment Resemble Hydrothermal Vent Taxa," *Science* 226, no. 4677 (1984): 965–967, https://doi.org/10.1126/science.226.4677.965.

47. See, for example, Ian R. Macdonald et al., "Chemosynthetic Mussels at a Brine-Filled Pockmark in the Northern Gulf of Mexico," *Science* 248, no. 4959 (1990): 1096–1099, https://doi.org/10.1126/science.248.4959.1096; and Bob Carney, "Lakes Within Oceans," National Oceanic and Atmospheric Administration, Ocean Explorer, last modified June 25, 2010, http://www.oceanexplorer.noaa.gov/explorations/02mexico/background/brinepool/brinepool.html.
48. Derek Sawyer et al., "Submarine Landslides Induce Massive Waves in Subsea Brine Pools," *Scientific Reports* 9, no. 128 (2019): 1–9, https://doi.org/10.1038/s41598-018-36781-7.
49. Craig R. Smith et al., "Vent Fauna on Whale Remains," *Nature* 341 (1989): 27–28, https://doi.org/10.1038/341027a0.
50. Steffen Kiel, "Whale and Wood Falls," in *Encyclopedia of Geobiology, Encyclopedia of Earth Sciences Series*, ed. Joachim Reitner and Volker Thiel (New York: Springer, Dordrecht, 2011), 901–904; Craig R. Smith et al., "Seven-Year Enrichment: Macrofaunal Succession in Deep-Sea Sediments Around a 30 Tonne Whale Fall in the Northeast Pacific," *Marine Ecology Progress Series* 515 (2014): 1–133, https://doi.org/10.3354/meps10955.
51. Amy R. Baco and Craig R. Smith, "High Species Richness in Deep-Sea Chemoautotrophic Whale Skeleton Communities," *Marine Ecology Progress Series* 260 (2003): 109–114, https://doi.org/10.3354/meps260109.
52. Christina Bienhold et al., "How Deep-Sea Wood Falls Sustain Chemosynthetic Life," *PLoS One* 8, no. 1 (2013): e53590, https://doi.org/10.1371/journal.pone.0053590.
53. Ruth D. Turner, "Wood-Boring Bivalves, Opportunistic Species in the Deep Sea," *Science* 180, no. 4093 (1973): 1377–1379, https://doi.org/10.1126/science.180.4093.1377.
54. Douglas Martin, "Ruth D. Turner, 85, Expert on the Wood-Eating Mollusk," *New York Times*, May 9, 2000, http://www.nytimes.com/2000/05/09/us/ruth-d-turner-85-expert-on-the-wood-eating-mollusk.html.
55. Angelo F. Bernardino et al., "Comparative Composition, Diversity and Trophic Ecology of Sediment Macrofauna at Vents, Seeps and Organic Falls," *PLoS One* 7, no. 4 (2012): e33515, https://doi.org/10.1371/journal.pone.0033515.
56. Bob Graham et al., *National Commission on the BP Deepwater Horizon Oil Spill and Offshore Drilling. Deep Water: The Gulf Oil Disaster and the Future of Offshore Drilling* (Washington, DC: United States Government Printing Office, 2011), 23.
57. For example, 1980 American Society for Civil Engineers Award to the Shell "Cognac" Project; see Graham et al., *National Commission on the BP Deepwater Horizon Oil Spill*, 31.
58. Graham et al., *National Commission on the BP Deepwater Horizon Oil Spill*, 63.
59. Bureau of Ocean Energy Management, Regulation and Enforcement, *Report Regarding the Causes of the April 20, 2010 Macondo Well Blowout* (Washington, DC: U.S. Department of the Interior, 2011), 34–35.
60. Martin Hovland, *Deep-Water Coral Reefs: Unique Biodiversity Hot-Spots* (New York: Springer, 2008), 1–304.
61. Fanny Girard and Charles R. Fisher, "Long-Term Impact of the Deepwater Horizon Oil Spill on Deep-Sea Corals Detected After Seven Years of Monitoring," *Biological Conservation* 225 (September 2018): 117–127, https://doi.org/10.1016/j.biocon.2018.06.028; "Deep-Sea Corals," Smithsonian Oceans, accessed April 28, 2023, https://ocean.si.edu/ecosystems/coral-reefs/deep-sea-corals.
62. Erik Cordes, "Why Explore the Deep Sea?" (presentation, TEDxTempleU, Philadelphia, PA, June 6, 2013), http://www.youtube.com/watch?v=J4aOhZNcW4I.

63. "History of *Alvin*."
64. "Mesophotic Habitats," National Oceanic and Atmospheric Administration, Flower Garden Banks, National Marine Sanctuary, accessed April 28, 2023, http://www.flowergarden.noaa.gov/about/mesophotic.html; Gregory A. Minnery, Richard Rezak, and Thomas J. Bright, "Depth Zonation and Growth Form of Crustose Coralline Algae: Flower Garden Banks, Northwestern Gulf of Mexico," in *Paleoalgology Contemporary Research and Applications*, ed. Donald F. Toomey and Matthew H. Nitecki (New York: Springer, 1985), 237–246.
65. "History," National Oceanic and Atmospheric Administration, Flower Garden Banks, National Marine Sanctuary, accessed April 28, 2023, http://www.flowergarden.noaa.gov/about/history.html.
66. "NOAA Deep-Sea Coral & Sponge Map Portal," National Oceanic and Atmospheric Administration, National Centers for Environmental Information, accessed April 28, 2021, http://www.ncei.noaa.gov/maps/deep-sea-corals/mapSites.htm.
67. Les Watling and Elliott A. Norse, "Disturbance of the Seabed by Mobile Fishing Gear: A Comparison to Forest Clearcutting," *Conservation Biology* 12, no. 6 (December 1998): 1180–1197, https://doi.org/10.1046/j.1523-1739.1998.0120061180.x.
68. Cordes, "Why Explore the Deep Sea?"
69. Bureau of Ocean Energy Management, Regulation and Enforcement, *Report Regarding the Causes of the April 20, 2010 Macondo Well Blowout* (Washington, DC: U.S. Department of the Interior, 2011), 20.
70. Bureau of Ocean Energy Management, *Report Regarding the Causes*, 1–4.
71. Graham et al., *National Commission on the BP Deepwater Horizon Oil Spill*, 114.
72. Graham et al., *National Commission on the BP Deepwater Horizon Oil Spill*.
73. Per, for example, the 1980 American Society for Civil Engineers Award to the Shell "Cognac" Project. See Graham et al., *National Commission on the BP Deepwater Horizon Oil Spill*, 146. The peak flow was approximately 60,000 barrels per day, or 2.52 million gallons: 2,520,000/24 = 105,000 gallons/hour = 1,750 gallons/minute.
74. Raffi Khatchadourian, "The Gulf War," *New Yorker*, March 6, 2011, http://www.newyorker.com/magazine/2011/03/14/the-gulf-war.
75. Samantha B. Joye et al., "The Gulf of Mexico Ecosystem, Six Years After the Macondo Oil Well Blowout," *Deep Sea Research Part II: Topical Studies in Oceanography* 129 (2016): 4–19, https://doi.org/10.1016/j.dsr2.2016.04.018.
76. National Commission on the BP Deepwater Horizon Oil Spill and Offshore Drilling. Deep Water: The Gulf Oil Disaster and the Future of Offshore Drilling (Washington, DC: U.S. Government Printing Office, 2011), 133. http://www.govinfo.gov/app/details/GPO-OILCOMMISSION.
77. Graham et al., *National Commission on the BP Deepwater Horizon Oil Spill*, 165–166.
78. Joye et al., "The Gulf of Mexico Ecosystem."
79. Helen K. White et al., "Impact of the Deepwater Horizon Oil Spill on a Deep-Water Coral Community in the Gulf of Mexico," *Proceedings of the National Academy of Sciences* 109, no. 50 (2012): 20303–20308, https://doi.org/10.1073/pnas.1118029109; Charles R. Fisher, Paul A. Montagna, and Tracey T. Sutton, "How Did the Deepwater Horizon Oil Spill Impact Deep-Sea Ecosystems?," *Oceanography* 29, no. 3 (2016): 182–195, https://doi:10.5670/oceanog.2016.82; Gregory S. Boland et al., "State of Deep-Sea Coral and Sponge

Ecosystems of the Gulf of Mexico Region: Texas to the Florida Straits," in *The State of Deep-Sea Coral and Sponge Ecosystems of the United States*, ed. Thomas F. Hourigan, Peter J. Etnoyer, and Stephen D. Cairns (Silver Spring, MD: National Oceanic and Atmospheric Administration, 2016), 1–59.
80. Girard and Fisher, "Long-Term Impact."
81. Patrick T. Schwing et al., "A Synthesis of Deep Benthic Faunal Impacts and Resilience Following the Deepwater Horizon Oil Spill," *Frontiers in Marine Science* 7 (2020): 560012, https://doi.org/10.3389/fmars.2020.560012.
82. Cordes, "Why Explore the Deep Sea?"
83. Girard and Fisher, "Long-Term Impact."
84. Schwing et al., "A Synthesis of Deep Benthic Faunal Impacts."
85. Marla M. Valentine, "Characterization of Epibenthic and Demersal Megafauna at Mississippi Canyon 252 Following the Deepwater Horizon Oil Spill" (master's thesis, Louisiana State University, 2013); Marla M. Valentine and Mark C. Benfield, "Characterization of Epibenthic and Demersal Megafauna at Mississippi Canyon 252 Shortly After the Deepwater Horizon Oil Spill," *Marine Pollution Bulletin* 77, no. 1–2 (2013): 196–209, https://doi.org/10.1016/j.marpolbul.2013.10.004.
86. Patrick T. Schwing et al., "A Decline in Benthic Foraminifera Following the Deepwater Horizon Event in the Northeastern Gulf of Mexico," *PLoS One* 10, no. 5 (2015): e0128505, https://doi.org/10.1371/journal.pone.0120565; Paul A. Montagna et al., "Deep-Sea Benthic Footprint of the Deepwater Horizon Blowout," *PLoS One* 8, no. 8 (2013): e70540, https://doi.org/10.1371/journal.pone.0070540.
87. Valentine and Benfield, "Characterization of Epibenthic and Demersal Megafauna."
88. Valentine and Benfield, "Characterization of Epibenthic and Demersal Megafauna."
89. Valentine, "Characterization of Epibenthic and Demersal Megafauna." See, for example, page 185: "When the near-absence of baseline information on the deep living communities is coupled with this spatiotemporal complexity, it is easy to grasp the inherent difficulty of assessing the impact of the DWH oil spill on GoM deep pelagic ecosystems." In 100 pages, the word "difficult" appears twelve times, the word "impossible" five times, the phrase "not possible" twice more.
90. John P. Incardona et al., "Deepwater Horizon Crude Oil Impacts the Developing Hearts of Large Predatory Pelagic Fish," *Proceedings of the National Academy of Sciences* 111, no. 15 (2014): E1510-E1518, https://doi.org/10.1073/pnas.1320950111.
91. Fisher et al., "Deepwater Horizon Oil Spill," 189.
92. Theodore W. Pietsch and Tracey T. Sutton, "A New Species of the Ceratioid Anglerfish Genus *Lasiognathus* Regan (Lophiiformes: Oneirodidae) from the Northern Gulf of Mexico," *Copeia* 103, no. 2 (2015): 429–432, https://doi.org/10.1643/CI-14-181.
93. Fisher et al., "Deepwater Horizon Oil Spill," 189.
94. Uta Passow and Edward B. Overton, "The Complexity of Spills: The Fate of the Deepwater Horizon Oil," *Annual Review of Marine Science* 13, no. 1 (2021): 109–136, https://doi.org/10.1146/annurev-marine-032320-095153.
95. Uta Passow and Scott A. Stout, "Character and Sedimentation of 'Lingering' Macondo Oil to the Deep-Sea After the Deepwater Horizon Oil Spill," *Marine Chemistry* 218 (2020): 103733, https://doi.org/10.1016/j.marchem.2019.103733.

96. Gregg R. Brooks et al., "Sedimentation Pulse in the NE Gulf of Mexico Following the 2010 DWH Blowout," *PLoS One* 10, no. 7 (2015): e0132341, https://doi.org/10.1371/journal.pone.0132341.
97. Passow and Stout, "Character and Sedimentation."
98. Fisher et al., "Deepwater Horizon Oil Spill," 189.
99. Craig R. McClain, Clifton Nunnally, and Mark C. Benfield, "Persistent and Substantial Impacts of the Deepwater Horizon Oil Spill on Deep-Sea Megafauna," *Royal Society Open Science* 6 (August 2019): 191164, https://doi.org/10.1098/rsos.191164.
100. "Lack of adequate sedimentation to cap oiled sediments was evident in ROV observations of a crab individual re-exposing oiled sediments as it walked." McClain, Nunnally, and Benfield, "Persistent and Substantial Impacts," 191164.
101. "History of *Alvin*."
102. Woods Hole Oceanographic Institution (WHOI) notes on this incident are insightful and funny, including "*Alvin* remained on the bottom until Labor Day [1969]. The DSV *Aluminaut* (a submersible from the Reynolds Aluminum Company) and the R/V *Mizar* assisted in the recovery, which required placement of a lifting bar into *Alvin*'s hatch (*Aluminaut* pilots had to break the sail in order to accomplish this). *Mizar* then raised *Alvin* to 50 feet, where divers wrapped the sub with lines and nets to prevent loss of any pieces. *Alvin* was towed to Martha's Vineyard, where a crane mounted on a barge pulled it out of the water. Overall, there was very little structural damage to the submersible (except for the sail). Lunches left on board were soggy but edible. Discovery that near-freezing temperatures and the lack of decaying oxygen at depth aided preservation opened up new areas of biological and chemical research. *Alvin* underwent a major overhaul after its ten-month dunking." "History of *Alvin*."
103. Peter Smith, "What Sunken Sandwiches Tell Us About the Future of Food Storage," *Smithsonian Magazine*, May 23, 2012, http://www.smithsonianmag.com/arts-culture/what-sunken-sandwiches-tell-us-about-the-future-of-food-storage-102837326/.
104. Beth N. Orcutt, "Living Between a Rock and Hard Place: Microbial Life in the Deep Sea and Potential Impacts of Deep-Sea Mining" (webinar, National Oceanic and Atmospheric Administration, National Marine Sanctuaries Webinar Series, March 17, 2022), https://sanctuaries.noaa.gov/education/teachers/living-between-a-rock-and-hard-place.html.
105. Diva J. Amon, "The Deep Ocean Is Essential for Life on Earth—But It Is Under Threat," Pew Charitable Trusts, March 2, 2022, http://www.pewtrusts.org/en/research-and-analysis/articles/2022/03/02/the-deep-ocean-is-essential-for-life-on-earth-but-it-is-under-threat.
106. "Luʻuaeaahikiikapapakū: Searching for Deep Sea Corals Among Ancient Volcanoes," Nautilus Live: Ocean Exploration Trust, http://www.nautiluslive.org/blog/2021/11/13/luuaeaahikiikapapaku-searching-deep-sea-corals-among-ancient-volcanoes.
107. "Never-Before-Surveyed Seamount Reveals Stunning Biodiversity," Nautilus Live: Ocean Exploration Trust, accessed April 28, 2023, nautiluslive.org/video/2021/12/03/never-surveyed-seamount-reveals-stunning-biodiversity.
108. "Unnamed Seamount A's Beautiful Bamboo Coral," Nautilus Live: Ocean Exploration Trust, accessed April 28, 2023, http://www.nautiluslive.org/video/2021/11/29/unnamed-seamount-beautiful-bamboo-coral.

109. Elizabeth M. De Santo, Elizabeth Mendenhall, and Elizabeth Nyman, "A Rush Is on to Mine the Deep Seabed, with Effects on Ocean Life That Aren't Well Understood," *Conversation*, August 17, 2020, http://www.theconversation.com/a-rush-is-on-to-mine-the-deep-seabed-with-effects-on-ocean-life-that-arent-well-understood-139833.
110. "About ISA," International Seabed Authority, accessed April 28, 2023, http://www.isa.org.jm/about-isa; De Santo, Mendenhall, and Nyman, "A Rush Is on to Mine the Deep Seabed."
111. Stefanie Kaiser, Craig R. Smith, and Pedro Martinez Arbizu, "Editorial: Biodiversity of the Clarion Clipperton Fracture Zone," *Marine Biodiversity* 47 (2017): 259–264, https://doi.org/10.1007/s12526-017-0733-0.
112. "Mapping the Clarion-Clipperton Fracture Zone," Nautilus Live: Ocean Exploration Trust, https://nautiluslive.org/blog/2018/10/01/mapping-clarion-clipperton-fracture-zone.
113. Kaiser, Smith, and Martinez Arbizu, "Editorial: Biodiversity of the Clarion."
114. Bart De Smet et al., "The Community Structure of Deep-Sea Macrofauna Associated with Polymetallic Nodules in the Eastern Part of the Clarion-Clipperton Fracture Zone," *Frontiers in Marine Science* 4 (2017): 103, https://doi.org/10.3389/fmars.2017.00103.
115. Jeremy Spearman et al., "Measurement and Modelling of Deep Sea Sediment Plumes and Implications for Deep Sea Mining," *Scientific Reports* 10, no. 1 (2020): 5075, https://doi.org/10.1038/s41598-020-61837-y.
116. Elizabeth Kolbert, "The Deep Sea Is Filled with Treasure, but It Comes at a Price," *New Yorker*, June 14, 2021, https://www.newyorker.com/magazine/2021/06/21/the-deep-sea-is-filled-with-treasure-but-it-comes-at-a-price.
117. Epeli Hau'ofa, "Our Sea of Islands," in *We Are the Ocean: Selected Works* (Honolulu: University of Hawai'i Press, 2008), 27–40.
118. Pacific Elders (@PacificElders), "@PacificElders call for refrain from #deepseabedmining and consideration of the #oceans spiritual significance for the #Pacific's #indigenous peoples," Twitter, February 24, 2022, http://www.twitter.com/PacificElders/status/1497098286867841029/photo/1.
119. Craig R. Smith et al., "Deep-Sea Misconceptions Cause Underestimation of Seabed-Mining Impacts," *Trends in Ecology and Evolution* 35, no. 10 (October 2020): 853–857, https://doi.org/10.1016/j.tree.2020.07.002.
120. Amon, "The Deep Ocean Is Essential."
121. Diva J. Amon et al., "Assessment of Scientific Gaps Related to the Effective Environmental Management of Deep-Seabed Mining," *Marine Policy* 138 (April 2022): 105006, https://doi.org/10.1016/j.marpol.2022.105006.

EPILOGUE

1. Alistair J. Hobday, Mia J. Tegner, and Peter L. Haaker, "Over-Exploitation of a Broadcast Spawning Marine Invertebrate: Decline of the White Abalone," *Reviews in Fish Biology and Fisheries* 10 (2000): 493–514, https://doi.org/10.1023/A:1012274101311.
2. National Marine Fisheries Service and National Oceanic and Atmospheric Administration, "Endangered and Threatened Species; Endangered Status for White Abalone," *Federal Register* 66, no. 103 (May 29, 2001): 29046–29055, https://www.govinfo.gov/content/pkg/FR-2001-05-29/pdf/01-13430.pdf#page=1.

3. Ryder Diaz, "At a Snail's Pace," Science Notes 2013, accessed May 1, 2023, https://sciencenotes.ucsc.edu/2013/pages/abalone/abalone.html.
4. Jacquelyn Ross, "No Word for Goodbye: Reclaiming Abalone's Home on the California Coast," Terralingua, November 11, 2020, https://terralingua.org/langscape_articles/no-word-for-goodbye-reclaiming-abalones-home-on-the-california-coast/.

INDEX

abalone: Pinto, 78; sea otters relation to, 72; shells of, 12–13; white, 209–210, 211–212, 213
Aboriginal Australians, 27
ACL. *See* Annual Catch Limit
acoustic gear triggers, 123
Acropora hyacinthus, 46–47
Acroporids, 26
Adelie penguins, 159
adze, 140
Africa, *Homo sapiens in*, 105
Agulhas Bank, South Africa, 126–127
Ahutoru, 137
AIS. *See* automated identification system
Alaska: Aleutian Islands, 60; Gulf of, 3
Albright, Rebecca, 26, 29, 42–43; *Acropora hyacinthus* and, 46–47; at One Tree Island, 27–28
Aleutian Islands, Alaska, 60
algae, 49–50; seagrass relation to, 57
algal blooms, 124; biodiversity affected by, 18; in Chukchi Sea, 163–164
alginate, 68
Alor, Indonesia, Tron Bon Lei, 107
Alvin (submersible), 188–190, 194, 201, 250n44, 254n102; in Santa Catalina Basin, 191
Amanthia carpenteri, 180
American Geophysical Union, 38
Amon, Diva, 202, 206
Amundsen, Roald, 161–162

anchoveta, 108
Ancient Sea Gardens (documentary), 80, 81
anemone, 16; sunburst, 20
Angus (submersible), 188
Annual Catch Limit (ACL), 235n52
Antarctica: in "Blue Marble" photograph, 149; ice in, 150–152; New Zealand relation to, 169–170; Rangiora in, 168–169; Ross Sea Region Marine Protected Area in, 102; Thwaites Glacier of, 153; Weddell seals in, 157–159
Antarctic Pack Ice Seal Program, 159
Anthopleura elegantissima, 16
Anthopleura sola, 16
Anthopleura xanthogrammica, 16
Aotearoa, New Zealand, 103, 138, 233n21
Apollo mission, 149
aquaculture, 89–90, 94; oyster, 92
Aquilino, Kristin, 211–212
Arctic bowhead whale, 162
Arctic Ocean, 150; Amundsen voyage of, 161–162; Atlantification of, 162–163; microplastics in, 168; polar ice in, 150–151, 160
Arctic Oscillation, 113
Arctic Research Consortium, 165
Aristotle, 2
Arizona, University of, 219n32
arthropods, 201
Asitau Kuru, 106, 234n32
Aspergillus sydowii, 24–25, 29

258   INDEX

Atlantic Ocean, 59; *Challenger* in, 180–181; data cables in, 185–186; map of, 182; NAO in, 113–114; North, 117–118; telegraph wires in, 179
Atlantification, of Arctic Ocean, 162–163
Atu, Te, 169
Auckland Islands, 169
Augustine, Skye, 88–89
Australia: Budj Bim Cultural Landscape, 100; colonization of, 133; Coral Sea Marine Park, 102; Great Barrier Reef Marine Park of, 36; heat waves in, 55; kelp forests of, 53, 59, 65–66, 69
Australians, Aboriginal, 27
automated identification system (AIS), 110
automation, for coral reef restoration, 48
autonomous underwater vehicles (AUVs), 174
Ayuda Foundation, 104

Ballard, Robert, 188–189, 202
Bangladesh, 156
Barents Sea, 163
Barry, Jim, 7, 8, 18; MBARI relation to, 173–174; in Pacific Grove, 16, 20
Baxter, Chuck, 7, 8
Beaufort Sea, 161
Beck Kehoe, Alice, 133, 134; on *Hōkūle'a*, 139
Belize, 47–48
Bell, Mauritius Valente, 176
Bell Laboratories, 183
beluga whales, 160–161
Benfield, Mark, 199, 201
Bering Sea, 164
Bermuda, 47
Between Pacific Tides (Ricketts), 6
Bigelow Laboratory for Ocean Science, 202
biodiversity, 103; in Agulhas Bank, 126–127; algal blooms effect on, 18; of coral reefs, 26, 28–29; of Haida Gwaii, 70; hydrothermal vents and, 202; krill and, 157; in Southern Ocean, 171; of tide pools, 4; of whale falls, 191
Biosphere 2 project, 18
black corals, 194
"Blob, The," 62

"Blue Marble" (photograph), 149, 170
boat technology, 107–108
Bodega Marine Laboratory, of University of California, Davis, 20, 211–212
Bonham, Charlton, 124
borealization, 163
Bouchard, Randy, 80–81
Bougainville, Louis-Antoine de, 137
Bound, Mensun, 171
BP oil, 193
Braverman, Irus, 23, 34, 43–44, 45, 47, 48
Briareaum asbestinum, 25
bridled terns, 27–28
brine pools, 190–191
Broughton Archipelago, 75
Brunet, Chris, 152
Budj Bim Cultural Landscape, Australia, 100
Byrne, Maria, 27, 30, 42

Calanus finmarchicus, 117; ocean temperature and, 119
Caldeira, Ken, 41, 42, 71
California: abalone in, 209–210; Cape Mendocino, 5; fisheries in, 111; kelp forests of, 55, 62; marine heat wave in, 20–21, 73, 124; Monterey Bay, 4–5, 173; ocean temperature in, 20–21; offshore oil platforms in, 193; Pacific Grove, 3–4, 5–7, 13–14, 16, 17, 20, 173; San Diego, 7, 144; San Francisco, 131; Snapshot Cal Coast in, 15; Tomales Bay, 92–93; whale ship strikes in, 124
California Academy of Sciences, 20, 175; lab-raised corals of, 46–47
California Current ecosystem, 121–122
Canada: cod moratorium of, 108–109
Canadian Arctic, 166–167; Indigenous people in, 168
Cannery Row (Steinbeck), 6
Cape Mendocino, California, 5
carbonate, 39, 40, 41
carbon dioxide, 39, 40; NOAA measuring, 38; ocean chemistry and, 90–91, 93–94; seaweed and, 94–95; in tide pools, 1–2
carbon sequestration, in seagrass, 57

Caribbean, 43; coral bleaching in, 33, 35; coral reefs in, 26, 29, 48–49; *Diadema antillarum* in, 49–50
Caribbean Coastal Marine Productivity (CARICOMP), 23
Carlton, James, 14
carp, 100
Carson, Rachel, 1, 2, 21, 192, 240n41
cartography, 136, 181
Catalina Island, 59–60
CCFZ. *See* Clarion-Clipperton Fracture Zone
Center for American Progress, 128
Challenger (English ship), 179–181
Chan, Francis, 125
Chasing Coral (film), 44
chemosynthetic environments, 190–192
Child, Julia, 109
Chile, kelp forests in, 68–69
Chukchi Sea, 160–161; algal blooms in, 163–164; Kivalina in, 166
Clam Garden Network, 88–89
clam gardens, 81–82, 95; fishponds compared to, 98, 100; in Pacific Northwest, 82–84, 87; of Salish Nation, 88; Salomon study of, 84–85
Clarion-Clipperton Fracture Zone (CCFZ), 205
climate change, 17; adaptation and restoration in response to, 41; Canadian Arctic and, 166; Caribbean and, 48; Chile affected by, 69; coastal hazards from, 166; cod affected by, 109; communicating to students about, 17; coral reefs affected by, 30, 36, 44; fisheries affected by, 127; Gulf Stream affected by, 118; Haida Nation affected by, 70–71; Hewatt transect affected by, 8–9; Intergovernmental Panel on, 39; mapping the ocean and, 186; ocean circulation, 118; ocean temperature and, 186; open-ocean animals and changes in distribution due to, 111–113; resource management relation to, 77; shellfish and, 94; Southern Ocean affected by, 159; whales affected by, 120–122, 124
Cloutier, Sheila Watt, 149

CNMI. *See* Commonwealth of the Northern Mariana Islands
coastal species, plastic pollution relation to, 146
Cobb, Kim, 37–38
cod, 108–109; industry and, 116
Cod (Kurlansky), 107, 109
Cold Spring Harbor Laboratory, New York, 6
Coleman, Melinda, 66
colonization, 70–71; of Australia, 133; of Society Islands, 240n18
Columbia (space shuttle), 140
Committee on Natural Resources, of U.S. House of Representatives, 38
Commonwealth of the Northern Mariana Islands (CNMI), 102–103
conservation, 72–73, 141; activism for, 128; COVID-19 relation to, 104; fisheries and, 109–110, 111, 112–113; in Hawai'i, 100–101; by Indigenous people, 85, 87, 103; laws for, 117; of North Atlantic right whale, 119–120; UN Convention on Biological Diversity for, 102; Western compared to Indigenous, 85
continental drift, 183
Cook, James, 137
coral-algal symbiosis, 28; coral bleaching and, 31
coral bleaching, 31, 32; in Caribbean, 33, 35; at Great Barrier Reef, 36, 53, 73; ocean temperature and, 33–34, 36; at Palmyra Atoll, 37
coral reefs, 193–194; algae relation to, 49–50; biodiversity of, 26, 28–29; *Briareaum asbestinum*, 25; carbonate relation to, 39; in Caribbean, 26, 29, 48–49; chemistry and, 40–41; climate change effect on, 30, 36, 44; diseases of, 23–25, 26, 35; El Niño events and, 34–36; fossils of, 32, 127, 154; ocean temperature and, 31, 35, 46, 48; at Palmyra Atoll, 37; reproduction of, 45–46; restoration of, 44–45, 47–48
coral research community, psychological effects on, 43–44
corals, deep-sea, 193–195; at Macondo, 198

Coral Sea Marine Park, Australia, 102
Coral Triangle, 26
Coral Whisperers (Braverman), 34, 43–44, 45; Gates in, 47; Hoegh-Guldberg in, 48
Cordes, Erik, 177, 189–190, 195; on coral reefs, 193–194; on Deepwater Horizon, 204; at Macondo, 198
Cousteau, Jacques, 183
COVID-19, 15, 94; conservation relation to, 104; supply chain shortages by, 154
crab fisheries, 122–124; low-oxygen waters effect on, 125
crabs, *Amanthia carpenteri*, 180
Craggs, Jamie, 46
Craven, Margaret, 76–77
crayweed, 66–67
crowdsourced data, 158–159
crown-of-thorns sea stars, 30
Currie, Jock, 126–127

Darwin, Charles, 2, 69
data cables, 185–186
DDT. *See* dichlorodiphenyltrichloroethane
"decade of the ocean," of United Nations, 185
deep sea, 189; coral in, 193–194, 198; microbes in, 201–202; mining in, 204–206; trash in, 176–177; Western scientists and, 178–179
Deepwater Horizon, 193, 195–197, 204, 253n89
deforestation, 233n21
Deur, Douglas, 82, 83, 85–87, 89–90
Diadema antillarum, 49–51
dichlorodiphenyltrichloroethane (DDT), 175
Dick, Adam, 76–77, 86; in Broughton Archipelago, 78; Deur relation to, 82, 85; on *loxiwey*, 80–81
Dillon, Rodney, 54, 55
dinoflagellate, 28, 32
Discovery Bay, Jamaica, 50
diseases: of coral reefs, 23–25, 26, 35; of oysters, 94; of sea stars, 63–64; of white abalone, 211
domoic acid, 62, 122
drought, Arctic temperatures relation to, 151
Dutton, Andrea, 153–155
Dzawada'enuxw Nation, 75–76

Earle, Sylvia, 54; at Catalina Island, 59–60
earthquakes, 183
East African Rift, 183
East Timor, 106
Eckstein, Lars, 136–137
ecosystems: California Current, 121–122; human evolution and, 239n9; around Mississippi Canyon Block 252, 199; of tide pools, 2–3
Ecuador, 143
Edward Ricketts State Marine Conservation Area, 4
El Niño events, 25, 32; coral reefs and, 34–36; kelp forests and, 55–56; weather changes and, 113
Emperor penguins, 159
End Triassic Extinction, 32
Endurance (English ship), 170–171
Endurance22 project, 171
entanglements, of whales, 120–122
environmental change, 19–20
Erickson, Kaare Sikuaq, 164–166, 167
Erlandson, Jon, 61
European Union, 102
E/V *Nautilus*, 202, 203; CCFZ mapped by, 205
Ewing, Maurice "Doc," 184
extreme heat, 18

factory ships, 108
farming: fish, 100; oyster, 92–93; shellfish, 95
Fears, Darryl, 54, 55
Ferrini, Vicki, 185
Finney, Ben, 138–139
First Nations people, 86; Clam Garden Network of, 88–89; clam gardens of, 85; Kwakwaka'wakw, 75–77, 78, 80–81, 82, 85; marine knowledge of, 78
Fish and Wildlife Service, U.S., 210
fisheries: abalone, 209; climate change effect on, 127; conservation and, 109–110, 111, 112–113; crab, 122–124, 125
fish farming, 100
fishhooks, 106–107

fishing industry, 104–105, 126; Huxley relation to, 128; plastic pollution relation to, 146–147; in Southern Ocean, 156–157; whales relation to, 117, 121
fishing trawls, 195
fishponds, clam gardens compared to, 98, 100
Florida Keys, coral bleaching in, 34
Flower Garden Banks, 195
food web: in Antarctic Ocean, 157, 159; in Arctic Ocean, 160; chemosynthetic, 189, 190, 192; heavy metals in, 168; on seafloor, 189; related to North Atlantic right whale, 120; West Coast crab fishey and, 122
Forbes, Edward, 178–179
fossil fuel companies: lawsuit against, 123; relationship to climate change, 45, 155; plastic pollution and, 146
fossils, of coral reefs, 154; coral bleaching and, 32; predation scars on, 127
Four Fish (Greenberg), 100, 104
Fresh Banana Leaves (Hernandez), 131, 218n22
funding, for deep sea exploration, 202

Galapagos Rift, 187–188
Gates, Ruth, 45, 47
Georges Bay, 109
Gies, Erica, 98
Gilman, Sarah, 8–9, 17, 23
Global Climate Action Summit, 131, 141–142
Global Fishing Watch, 110
global warming, 7, 9; deep sea affected by, 177; sea level relation to, 155. *See also* climate change
Glynn, Peter, 30–31, 33
Gore, Al, 17
Grand Banks, 107
gray whales, 121; Pacific, 162
Great Australian Bight, 65
Great Barrier Reef, 26, 45, 46; coral bleaching at, 36, 53, 73; Lizard Island in, 33–34; One Tree Island in, 27–28, 30, 42–43
Great Barrier Reef Marine Park, Australia, 36
Great Pacific Garbage Patch, 143–144, 146; Rochman at, 144–145

Greenberg, Paul, 100, 104
Greene, Charles, 118
Greenland, 155
Groesbeck, Amy, 84–85
Guam, 104
Gulf of Alaska, 3
Gulf of Maine: mussels in, 14; ocean temperature in, 118
Gulf of Mexico, 192; chemosynthetic environment in, 190; deep-sea corals in, 194; Deepwater Horizon in, 193, 195–197, 204, 253n89; Macondo in, 193, 195–196, 197, 198–200, 201
Gulf of St. Lawrence, 120
Gulf Stream, 117; climate change effect on, 118
Gumbs, Alexis Pauline, 97, 115, 116
Gunditjmara people, 100

Haida Gwaii, 69–70; sea otters of, 71–72
Haida Nation, 70; marine knowledge of, 71
halocline, 162
Harari, Yuval Noah, 133
Hardy, Penelope, 115
Harper, John, 80–81
Harrington, John Peabody, 10
Harvell, Drew, 23–24, 33; on coral bleaching, 35; on coral research community, 43; on ocean conditions, 78–79; at San Salvador transect, 25; sea star research of, 63; at Uva Island, 30
Hau'ofa, Epeli, 137–138, 205
Hauser, Donna, 160–161, 162
Hawai'i, 241n29; black corals in, 194; conservation in, 100–101; Indigenous people of, 72; Kāne'ohe Bay in, 45; Kaua'i, 98; Mauna Kea volcano in, 140; navigation to, 135; Papahānaumokuākea National Monument in, 102, 128, 178, 203
Hawaiki, 138
heat waves, in Australia, 55
heavy metals, 168
He'eia estuary, pollution of, 99
He'eia NERR, 99, 105
He'eia Stream, 98–99

262 INDEX

Heezen, Bruce, 182–184, 185
Hernandez, Jessica, 131, 218n22
herring, Pacific, 87–88
Hewatt, Willis G., 5–7
Hewatt transect, 6, 7–8, 173; climate change effect on, 8–9; Macondo compared to, 199–200; Micheli at, 19; San Salvador transect compared to, 25
Hikianalia (canoe), 131, 141, 147
H. neanderthalensis, 79
Hodin, Jason, 64
Hoegh-Guldberg, Ove, 33–34, 35, 48
Hog Island Oyster Company, 89, 90, 93
Hōkūleʻa (canoe), 139–140, 141, 147
Homo erectus, 79, 132
Homo sapiens, 105, 133
Hopkins Marine Station, Stanford University, 4, 6
Hopkins rose nudibranchs, 16
Hughes, Terry, 36
human evolution, 239n9
humpback whales, 121
hurricanes, coral bleaching and, 33
Huxley, Thomas, 101, 104–105, 107; fishing industry relation to, 128
hydrothermal vents, 184, 189; biodiversity and, 202; whale falls compared to, 191
hypoxic zone, 125

I Heard the Owl Call My Name (Craven), 76–77
Ikaaġun, 165
iNaturalist, 15
Indigenous science, 87, 218n22; Western science and, 73, 77
Indonesia, Alor, 107
industry: cod and, 116; fishing, 104–105, 117, 121, 126, 146–147, 156–157; heavy metals from, 168; NGO *versus*, 101
Institute for Socio-Ecological Research (ISER), 50
Intergovernmental Panel on Climate Change (IPCC), 39
International Convention for the Safety of Life at Sea (SOLAS), 110
International Seabed Authority, 204

International Union for Conservation of Nature (IUCN), 64, 128
Iñupiat people, 164–165
invasive species, 5
IPCC. *See* Intergovernmental Panel on Climate Change
ISER. *See* Institute for Socio-Ecological Research
Islanders (Thomas), 135
IUCN. *See* International Union for Conservation of Nature

Jahren, Hope, 75, 156, 209
Jamaica, Discovery Bay, 50
Japan, 106, 133
Java Trench, 132
jewelry, of Ohlone people, 11, 12
Johannes, Robert, 103
Johnson, Rebecca, 15–16
Johnson Sea Link (deep-sea submersible), 92
Juárez, Benito, 115

Kamalu, Lehua, 131–132, 141
Kamchatka Peninsula, Russia, 60
Kanahele-Mossman, Huihui, 178
Kane, Haunani, 141
Kāne, Herb Kawainui, 138–139
Kāneʻohe Bay, in Hawaiʻi, 45
Kauaʻi, Hawaiʻi, 98
Kealaikahiki, 135
kelp forests, 61, 73; of Australia, 53, 59, 65–66, 69; of California, 55, 62; in Chile, 68–69; ocean temperature and, 54–55, 59; pollution relation to, 60; restoration of, 66–67; seagrass compared to, 57, 58; shellfish compared to, 94; urchins relation to, 55–56, 64, 71
Kennedy, Dorothy, 80–81
"keystone species," 2
Khatchadourian, Raffi, 197
killer whales, 78
Kim, Kiho, 24–25, 35
Kingcome River, 75, 76–77
Kivalina, 166
Kleypas, Joanie, 39–42, 91

Korea, 64
krill oil, 156–157
Krumhansl, Kira, 58–59; Pérez Matus relation to, 68–69
Kunghit Island, 69–70
Kurashima, Natalie, 100
Kurlansky, Mark, 107–108, 109
Kwakwaka'wakw Nation, 75–77, 85; *loxiwey* of, 80–81, 82; marine knowledge of, 78
Kwaxsistalla, 76, 79–80, 84, 85

Labe, Zack, 151
lab-raised corals, of California Academy of Sciences, 46–47
Lamont Geological Laboratory, 181
Langscape (magazine), 212
LaRue, Michelle, 157–159
laws, for conservation, 117
lawsuit, against fossil fuel companies, 123
Learning from the Octopus (Sagarin), 17–18
Leon Guerrero, Carlotta, 104
Lepofsky, Dana, 79, 81–82, 86, 87, 94
Lessonia, 68
Liboiron, Max, 145–146
Limahuli Garden and Preserve, 98
Lizard Island, in Great Barrier Reef, 33–34
Lophelia pertusa, 194, 198
"Lost and Found" (essay), 10
Louisiana, 152
Lovers Point–Julia Platt State Marine Reserve, 4
low-oxygen waters, 124–125
loxiwey, 80–81, 82
Lubchenco, Jane, 2, 90
Lu'uaeaahikiikapapakū, 203

Macondo, 193, 195–196, 197; arthropods at, 201; deep-sea corals at, 198; Hewatt transect compared to, 199–200
Macrocystis, 68
Mahirta, 107
Maine: Gulf of, 14, 118; sea level in, 152
Makah people, 72
Māori people, 168–169; in Southern Ocean, 170

Marean, Curtis, 3
Mariana Trench, 175
Mariana Trench Marine National Monument, 128
marine heat wave: in Australia, 65; in California, 20–21, 73, 124; El Niño events and, 25; kelp forests affected by, 55; in Pacific Ocean, 62, 122
Marine Isotope Stage 5e, 155
marine knowledge: of Haida Nation, 71; of Kwakwaka'wakw Nation, 78
Marine Mammal Commission, U.S., 120
marine protected areas (MPAs), 103–104, 128, 141
marine snow, 200; in CCFZ, 205
Márquez, Gabriel García, 193
Marquinez, L. Frank, 213
Mars (offshore oil platform), 193
Mathews, Darcy, 61
Mauna Kea volcano, 140
Maury, Matthew Fontaine, 114–115
MBARI. *See* Monterey Bay Aquarium Research Institute
McClain, Craig, 201
McMurdo Station, 157
Meadows, Isabel, 9–10, 11, 12, 77; Dick compared to, 85–86
Mediterranean: black corals in, 194; tide pools of, 2
Med Tapei (cargo ship), 175
Melanesia, 133–134
mele, 178
meridional overturning circulation, 118
Meso-American Reef, 25–26
mesophotic coral habitats, 176
Mexico, 115
Miami Beach, 152
Micheli, Fiorenza, 18–19
microbes, in deep sea, 201–202
microplastics, 145; in Arctic, 168
Mid-Atlantic Ridge, 182–183
middens, 79–80
migration, of whales, 115, 119–120, 124, 161
minerals, on seafloor, 203–204
mining, deep-sea, 204–206

Miranda, Deborah, 10
Mississippi Canyon, 193; Block 252, 197, 199; Block 294, 198
Miwok Tribe, 212–213
Moananuiākea, 147
monkfish, 109
Monterey Bay, California, 4–5, 173
Monterey Bay Aquarium, 7
Monterey Bay Aquarium Research Institute (MBARI), 173–174
Monterey Bay National Marine Sanctuary, 175
Monterey Submarine Canyon, 173
Moore, Michael, 116–117, 119; on acoustic gear triggers, 123; on entanglements, 121
moratorium: on cod, 108–109; on whaling, 156
Morris, Mary, 80
Morrison, Sandy, 170
mortality rate: of North Atlantic right whales, 120; of seals, 163
MPAs. See marine protected areas
mussels: in Gulf of Maine, 14; sea stars relation to, 18
Muwekma Ohlone tribe, 131

NAO. See North Atlantic Oscillation
National Estuarine Research Reserve (NERR), 98; He'eia in, 99, 105
National Geographic (magazine), 54, 59, 139, 186–187
National Indigenous and Earth Sciences Convergence Hub, 167
National Marine Fisheries Service (NMFS), 235n52
National Marine Sanctuary, U.S., 195
National Oceanographic and Atmospheric Administration (NOAA), 2, 38, 90, 117, 178; ACL regulated by, 235n52; fines to, 175; funding from, 202; North Atlantic right whales and, 119–120; seal mortality monitored by, 163; UME's tracked by, 237n87
National Science Foundation, U.S. (NSF), 164–165, 202, 246n50

navigation, 114–115, 131–132; Polynesian, 136–138, 142; of Polynesian Islands, 135–136
Navy, U.S., 114–115, 181, 187–188
NERR. See National Estuarine Research Reserve
New York, Cold Spring Harbor Laboratory, 6
New Zealand: Antarctica relation to, 169–170; Aotearoa, 103, 138, 233n21; urchins in, 73
NGO. See nongovernmental organization
Niagara, USS, 179
Nielsen, Erica, 20
Nielsen, Karina, 2, 19, 61, 62–63
Nijhuis, Michelle, 46
NMFS. See National Marine Fisheries Service
NOAA. See National Oceanographic and Atmospheric Administration
nongovernmental organization (NGO), 101, 110
North, Wheeler, 55–56
North Atlantic Drift, 117–118
North Atlantic Ocean, 117–118
North Atlantic Oscillation (NAO), 113–114
North Atlantic right whale, 115–117, 119–120, 162
North Atlantic Right Whale Consortium, 119
Norway, Porsanger Fjord, 64
NSF. See National Science Foundation, U.S.
Nunnally, Clifton, 201

ocean acidification, 41, 42, 94; in deep sea, 177; oysters and, 91–92, 93; seagrass relation to, 57
ocean chemistry: carbon dioxide and, 90–91, 93–94; coral reefs and, 40–41; in tide pools, 1–2
Ocean Exploration Trust, 202
ocean-floor maps, 181–182, 183–185
oceanic oscillations, 113–114
Ocean Modeling Forum, 87
Ocean Outbreak (Harvell), ocean conditions in, 25, 35, 78–79
ocean temperature: *Calanus finmarchicus and*, 119; in California, 20–21; *Challenger* measuring, 180; in Chile, 68; climate

change and, 186; conservation relation to, 73; coral bleaching and, 33–34, 36; coral reefs and, 31, 35, 46, 48; Gilman study of, 17; kelp forests and, 54–55, 59; NAO and, 113–114; in North Atlantic Ocean, 118; oysters relation to, 94
O'Connor, Sue, 106–107, 234n32
O'Donnell, Brady, 120
offshore oil platform, 192–193. *See also* Deepwater Horizon
Ogg, Dick, 123–124
Ohlone jewelry, 11, 12
Ohlone people, 4, 9–10, 11–12
'Ohu Gon, Sam, 99
Olivella shells, 11, 12
One Hundred Years of Solitude (Márquez), 193
One Tree Island, in Great Barrier Reef, 27–28, 30, 42, 43
On the Origin of Species (Darwin), 2
open-ocean animals, climate change and, 111–113
Operation Crayweed, 67
Orcutt, Beth, 202, 203
Oregon, low-oxygen waters in, 125
Our Sea of Islands (Hau'ofa), 138
oysters, 90, 91–92, 93, 94

Pacification, 163
Pacific Biological Laboratories, 6
Pacific Coast, 79–80
Pacific Coast Federation of Fishermen's Associations, 123
Pacific Decadal Oscillation, 113
Pacific Elders Voice, 205
Pacific Grove, California, 3–4, 173; Barry in, 16, 20; Hewatt at, 5–7; tide pool in, 13–14, 17
Pacific herring, 87–88
Pacific Islanders, 102–103
Pacific Northwest, clam gardens in, 82–84, 87
Pacific Ocean, 48, 142–143; data cables in, 185–186; marine heat wave in, 62, 122
Pacific sardine, 108
Paepae o He'eia, 99

Palau National Marine Sanctuary, 104
Palmyra Atoll, 37
Panama, Uva Island, 30–31
Papahānaumokuākea National Monument, 102, 128, 178; E/V *Nautilus* exploring, 203
Papee'ete, Tahiti, 139
parrotfish, 49–50
Partnership for Interdisciplinary Studies of the Coastal Ocean (PISCO), 125
passive acoustic recording, 120
Passow, Uta, 200
Patagonia, 69
Paull, Charlie, 166–167, 190
Pérez Matus, Alejandro, 67, 68–69
permafrost, 166–167
Pew Charitable Trusts, 128
Phoebe Hearst Museum, University of California, Berkeley, 11
phytoplankton, 159, 160
Piailug, Mau, 139–140, 141
Pinnacle Point, South Africa, 3, 79
Pinsky, Malin, 111–113, 126, 163
Pinto abalone, 78
PISCO. *See* Partnership for Interdisciplinary Studies of the Coastal Ocean
Pitcairn Islands, 102
plankton, 62, 150
plastic pollution, 143–145, 146–147
Poe, Melissa, 78, 79, 87–88
Point Cabrillo, California, 4
polar ice, 150–152; in Arctic Ocean, 160; Weddell seals and, 157–159
pollution, 101; coral bleaching relation to, 33; of He'eia estuary, 99; kelp forests relation to, 60; plastic, 143–145, 146–147; sea level and, 153–154
Polynesian Islands, navigation of, 135–136
Polynesian navigation, 136–138, 142
Polynesian Voyaging Society, 139, 141, 147
Pomo Tribe, 212–213
Porcupine, HMS, 179
Porsanger Fjord, Norway, 64
potlatch gatherings, 76
predation scars, 127
Pseudo-nitzschia, 122

psychological effects, on coral research community, 43–44
Puerto Rico, coral bleaching in, 33
Puerto Rico Trench, 177, 190
purple sea fan, 23

Quadra Island, 82, 84

Rangiora, Hui Te, 168–169
Recalma-Clutesi, Kim, 76, 77, 78, 80–81, 82, 86; on clam gardens, 84
remotely operated vehicle (ROV), 173–175, 195, 203; at Macondo, 198, 199
reproduction, of coral reefs, 45–46
resource management, climate change relation to, 77
restoration: of coral reefs, 44–45, 47–48; of kelp forests, 66–67
Rhode Island, 152–153
Ricketts, Edward, 6
Right to Be Cold, The (Cloutier), 149
Rising (Rush), 53, 152–153
Rocha, Luiz, 175–177
Rochman, Chelsea, 144–145, 146, 168
Romans, oyster aquaculture, 79
Rose Garden, 189
Ross, Jacquelyn, 212
Ross Sea, 157–158
Ross Sea Region Marine Protected Area, Antarctica, 102
ROV. *See* remotely operated vehicle
Rozwadowski, Helen, 115
Ruesink, Jennifer, 142, 143–144
Rumsen language, 10
Rush, Elizabeth, 53, 152–153
Russia, Kamchatka Peninsula, 60
R/V *Falkor*, 202–203
R/V *Mizar*, 254n102
Ryuku Islands, 133

Saba, island of, 23
Sagarin, Raphael, 8, 17–18, 23, 219n32
Saipan, 127–128, 129
Sakitari Cave, Japan, 106
Salish Nation, clam gardens of, 88

Salish Sea, 78
Salmond, Anne, 135–136
Salomon, Anne, 72, 84–85, 87, 88–89, 95
Samhouri, Jameal, 123
Samoa, 134
San Diego, California, 7, 144
Sanford, Eric, 20
San Francisco, California, 131
San Francisco Bay, 5, 15–16
San Salvador transect, 25
Santa Catalina Basin, 191
Santa Cruz Mountains, 173
sardine, Pacific, 108
Sargasso, 143
Sawyer, Terry, 90–91, 92–93
Schmidt Ocean Institute, 202–203
Schneider, Tsim, 5
Schwarz, Anja, 136–137
Scripps Institution of Oceanography, in San Diego, 7, 144
scuba diving, 23, 175–176
Sea Around Us (Carson), 192
Seabed 2030, 185
seabirds, of Chukchi Sea, 163
sea eagles, 27
sea fans, 23, 24–25
seafloor, 189, 203–204
seagrass, 56; kelp forests compared to, 57, 58
sea level, 152, 153–154, 155
sea lions, 62
seals: mortality rate of, 163; Weddell, 157–159
Sea of Cortez, 18
Sea of Glass, A (Harvell), 30
sea otters: of Haida Gwaii, 71–72; seagrass relation to, 57
sea stars, 17; crown-of-thorns, 30; diseases of, 63–64; mussels relation to, 18
seaweed, carbon dioxide and, 94–95
SECORE. *See* Sexual Coral Reproduction
sequoias, kelp forests compared to, 55
sewage, in Sydney Harbor, 66–67
Sewall, Laura, 152
Sexual Coral Reproduction (SECORE), 47
Seychelles, 154, 155
SFU. *See* Simon Fraser University

Sheffield, Gay, 164
shellfish, 79, 86, 230n17; in clam gardens, 82; farming of, 95; kelp forests compared to, 94; ocean acidification and, 92
Shell Oil Company, 192–193
"shifting baselines," 78
ship strikes, of whales, 124, 162
shipworms, 191–192
Shukla, Priya, 89, 90, 94, 124
Silent Spring (Carson), 2
Simon Fraser University (SFU), 79, 81–82
sixgill shark, 176
Skidegate Inlet (*Xaana Kaahlii*), 71
slave trade, 115
Smith, Garriet, 24
Smith, Percy, 168–169
Smithsonian Tropical Research Institute, 30
Snapshot Cal Coast, 15
Society Islands, 134, 240n18
SOLAS. *See* International Convention for the Safety of Life at Sea
sonar technology, 181, 182, 187
Sones, Jackie, 20
Sorte, Cascade, 14–15
South Africa: Agulhas Bank, 126–127; Pinnacle Point, 3, 79
South China Sea, 32
Southern Ocean, 150; biodiversity in, 171; climate change effect on, 159; fishing industry in, 156–157; Māori people in, 170
South Pacific, 132
Spanish settlers, Indigenous people relation to, 5
Spilhaus projection, 186
spiny lobster, 15–16, 65
Stanford University, Hopkins Marine Station of, 4, 6
starvation, 163
Steinbeck, John, 2, 6
Stevens, Michael, 169
stony corals, 31–32
Story of More, The (Jahren), 75, 156, 209
summer flounder, 112, 235n52
sunburst anemone, 20
Sunda Islands, 132–133

Sunda Plate, 132
sunflower stars, 64
supply chain shortages, 154
swordfish, 111
Syal, Richa, 156
Sydney Harbor, 66–67
symbiosis, coral-algal, 28

Tahiti, 134, 137, 139, 147, 241n29
Tasmania, 65, 68
technology: boat, 107–108; of crab fisheries, 123–124; fishhook, 106–107; mining, 204; sonar, 181, 182, 187; telegraph wire, 179
telegraph wires, in Atlantic Ocean, 179
Tharp, Marie, 181–182, 183–185, 187
Thomas, Nicholas, 135
Thompson, Nainoa, 131–132, 140, 141–142, 146
Thwaites Glacier, 153
tide pools, 1, 61, 62–63; anemone in, 16; biodiversity of, 4; ecosystems of, 2–3; oceans compared to, 19; Pacific Grove, 13–14, 17
Titanic shipwreck, 192
Tomales Bay, California, 92–93
Tongva-Acjachmemen Tribe, 213
Transocean, 196
trash, 176–177
Triassic period, 31–32
Tron Bon Lei, Alor, 107
tube worms, 189
tule canoes, 11
tuna, 104
Tupaia, 137–138
Turner, Nancy, 61, 86
Turner, Ruth, 191–192
Twenty Thousand Leagues Under the Sea (Verne), 178, 179

UCLA. *See* University of California, Los Angeles
Ukpeaġvik Iñupiat Corporation, 164
UME's. *See* Unusual Mortality Events
Unalakleet, 167
UN Convention on Biological Diversity, 102
Undersea (Carson), 1
Undrowned (Gumbs), 97

United Kingdom, 102
United Nations, "decade of the ocean" of, 185
United States (U.S.), 102; cod moratorium of, 108–109; Fish and Wildlife Service of, 210; House of Representatives of, 38; Marine Mammal Commission of, 120; National Marine Sanctuaries of, 195; Navy of, 114–115, 181, 187–188; NSF of, 164–165, 202, 246n50; sea level in, 152
University of Arizona, 219n32
University of California, Berkeley, Phoebe Hearst Museum of, 11
University of California, Davis, 89; Bodega Marine Laboratory of, 20, 211
University of California, Los Angeles (UCLA), 33
University of Hawai'i, 97
University of Puerto Rico, Mayagüez, 50
University of South Carolina Aiken, 24
Unusual Mortality Events (UME's), 237n87
urchins, 65, 73; *Diadema antillarum*, 49–51; kelp forests relation to, 55–56, 64, 71
U.S. *See* United States
U.S. House of Representatives, Committee on Natural Resources of, 38
Uva Island, Panama, 30–31

Valentine, Marla, 199
Van Parijs, Sofie, 120
Veach, Charles Lacy, 140
Ventana (ROV), 173–175
Vergés, Adriana, 55, 67
Verne, Jules, 178, 179
Villagomez, Angelo, 102–103, 127–129
Viosca Knoll 906, 198
Vizcaíno, Sebastián, 4–5
volcanoes, 31–32, 140

Walkes, Sam, 20
Wallace, Alfred Russel, 133
Washington State, 78
Waters, Suki, 213
We Are All Whalers (Moore), 119
weather changes, 186; El Niño events and, 113; polar ice and, 152

Weber, Michael, 109
Weddell Sea, 170–171
Weddell seals, 157–159
Wehi, Priscilla, 169–170
Wentworth Seamounts, 203
Western conservation, 85
Western explorers, 136
Western science, Indigenous science and, 73, 77
Western scientists, 115; deep sea and, 178–179; Indigenous people relation to, 72–73, 77, 105, 164–165
whale falls, 191
whales: beluga, 160–161; climate change effect on, 120, 122; crab fisheries relation to, 123; entanglements of, 120–122; fishing industry relation to, 117, 121; killer, 78; North Atlantic right, 115–117, 119–120, 162; Pacific gray, 121, 162; ship strikes of, 124, 162
whaling, moratorium on, 156
Whiskey Creek Hatchery, 90–91
white abalone, 209–210, 211–212, 213
WHOI. *See* Woods Hole Oceanographic Institute
Wickett, Michael, 41
Williams, Madi, 103, 233n21
Williams, Stacey, 50
Wilson, Kii'iljuus Barbara, 70–71, 73, 85
Winter, Kawika, 97, 99, 103–104, 105, 115, 129; on conservation, 100–101, 112
Woods Hole Oceanographic Institute (WHOI), 6, 116, 188
World Ocean Floor Map, 184
World War II, 107–108
Wright, Dawn, 185; in *National Geographic*, 186–187

Xaana Kaahlii (Skidegate Inlet), 71

Yamane, Linda, 10, 11–13, 19
Yang Yu Jing, 100
Young, Alison, 15–16

zooplankton, 160